Food Irradiation Now

NUTRITION SCIENCES

In preparation:
Bodwell CE and Petit L, eds.: Plant Proteins for Human Food

Food Irradiation Now

Proceedings of a Symposium,
held in Ede, the Netherlands, 21 October 1981

Sponsored by GAMMASTER, Ede, The Netherlands

1982
Martinus Nijhoff / Dr W. Junk Publishers
The Hague / Boston / London

Distributors:

for the United States and Canada

Kluwer Boston, Inc.
190 Old Derby Street
Hingham, MA 02043
USA

for all other countries

Kluwer Academic Publishers Group
Distribution Center
P.O. Box 322
3300 AH Dordrecht
The Netherlands

Library of Congress Cataloging in Publication Data
Main entry under title:

Food irradiation now.

  (Nutrition sciences ; v. 1)
  1. Food, Irradiated--Congresses. 2. Radiation
preservation of food--Congresses. 3. Food--
Effect of radiation on--Congresses. I. Series.
TX571.R3F63 1982   363.1'92   82-8164
AACR2
ISBN-13: 978-94-009-7620-7   e-ISBN-13: 978-94-009-7618-4
DOI: 10.1007/978-94-009-7618-4

Contents

Foreword

E.H. Kampelmacher (chairman)

The application of ionizing radiation is the most modern
technique being used in the battle against bacterial decay
and in the elimination of pathogenic microorganisms in foodstuffs. Intensive world-wide studies have shown that this
technique is effective, has no detrimental effects on human
health and can be applied safely.

These facts have all been confirmed recently by a Committee
of experts from the World Health Organization, who have been
carefully evaluating data since 1960. In spite of these
convincing data, most countries do not make use of the technique
at all, and some only hesitatingly. This is mainly due to the
emotional resistance of the consumer.

From the start the Netherlands has made a very important
contribution to the irradiation of food, through microbiological
and toxicological research as well as through the setting-up
of a pilot plant by the government and through the practical
application of "Gammaster" on a commercial basis. The initiative
of "Gammaster" to celebrate its tenth anniversary by organizing
a symposium in which all aspects of food irradiation are
considered is most laudable.

All the more so because it will help to remove the many
misunderstandings, will offer information and will indicate
to the potential user a method that can make an important
contribution to the prevention of decay and spoilage of foodstuffs and to the exclusion of food-borne infections and food
poisoning in man.

Introduction

J.G. Leemhorst

   The title "Food irradiation now" was intended as an
indication of the relevance of the theme of this proceedings.
But food irradiation is no novelty. Twenty years ago its
importance was recognized by UN organizations. During a meeting
of representatives of the IAEA, FAO and WHO, which took place
from 23 to 30 October 1961, it was decided to make an extensive
study of this method of treatment.

   For nineteen years, research institutes throughout the
world have intensively studied the effects of radiation on
foodstuffs. This period of research came to an end in November
1980. A Committee of experts, convened by the same UN organizations, declared that food irradiation was safe. The member
states of the UN were advised to permit the irradiation of
food. This advice led to the organization of a symposium on
21 October 1981 and the publication of this proceedings.

   As a division of the Cooperative Pharmaceutical Chemists
Association "De Onderlinge Pharmaceutische Groothandel" UA
(OPG) (The Associated Pharmaceutical Wholesalers), Gammaster's
main aim is to contribute towards the improvement of public
health. We believe that the proceedings is suited to this aim.

   The elimination of disease-producing microorganisms is the
most important application of food irradiation. Application
for the prevention of spoilage and decay could even help to
relieve the world's food problem.

   We are delighted that internationally recognized experts
were prepared to talk and write about the most important aspects
of this process. We are, therefore, confident that this
proceedings will contribute towards a greater appreciation
of the possibilities offered by this technique.

Whither protection of the consumer against enteropathogenic bacteria on fresh meats and poultry by processing for safety

D.A.A. Mossel and P. van Netten

Savoir pour prévoir, prévoir pour
prévenir. A. Comte (1798-1857).

1. INTRODUCTION

The consumer and the bacteriologist alike are periodically
alarmed by reports in the daily press concerning outbreaks of
food-transmitted, febrile gastroenteritis, attacking either
large numbers of victims, or leading to the deaths of elderly
people or susceptible patients. Food technology journals
often report that the treatment of foods with gamma rays is
an efficient tool for the elimination of non-sporing pathogens
from food and hence for the control of outbreaks of illness
due to enteric bacteria. There seems to be a contradiction
between these two series of facts. This prompts the following
questions:
(i) do outbreaks of enteric disease transmitted by foods
occur as frequently as is often suggested and what is their
impact on public health in general?
(ii) the irradiation of foods may indeed be effective for the
elimination of pathogens, but is it also acceptable from the
point of view of the health of the consumer?
(iii) if both answers tend to be in the affirmative, why then
is food irradiation not used at present for the specific purpose of controlling food-transmitted enteritis.
    This paper will deal with these questions. It will try to
provide an unbiased answer, based mainly on literature data
and partly on the senior author's own research on food
irradiation carried out since about 1960 (21, 23, 24).

Food Irradiation Now – ISBN 90-247-2703-0
Copyright Gammaster and Martinus Nijhoff/Dr W. Junk Publishers

2. MORBIDITY, AETIOLOGY AND IMPACT OF FOOD INFECTIONS

2.1. Epidemiological data

It is universally accepted that severe under-reporting of infectious diseases in general and of food-transmitted infections in particular occurs even in countries with an advanced Public Health infrastructure (18). An accurate estimate of the extent of under-reporting in The Netherlands can be derived from investigations carried out in Rotterdam. Practising physicians collected data on all outbreaks of diarrhoea occurring amongst their patients and their data were subsequently compared with the official, nation-wide reports (14). It has become clear from this survey that, as elsewhere in North-Western Europe and the United States, the numbers of persons really affected amount to 10 to 25 times the recorded incidence.

The causes of this marked under-reporting have also been revealed. Only approximately one third of the patients will consult a doctor at all. And, in turn, only about one third of the physicians consulted will ever send a specimen to a bacteriological laboratory. It is indeed striking that suffering from febrile gastroenteritis is nowadays considered to be so normal that victims do not seem to bother to call for medical attention.

Research has also been carried out to assess more accurately the aetiology of food-transmitted symptoms. This relies on attempts to identify systematically the foods involved in outbreaks. Procuring such specimens is not always easy. A prerequisite is that such left-overs are 'epidemiologically valid', i.e. that their microbiological community structure is not essentially different from that prevailing at the moment of consumption. Nonetheless, sufficient samples have been collected and examined, particularly in the region of Haarlem, to allow tentative conclusions to be drawn (3). The supplementary data obtained from these surveys have contributed to a better understanding of the aetiology of the incidence of food-transmitted diseases of microbial origin in The Netherlands.

2.2. Foods and mechanisms involved

The morbidity data referred to in the previous section do not essentially differ from what is observed in the USA and in North-Western Europe generally (4, 2). The main pathogens involved are (i) Gram negative, rod-shaped bacteria, particularly Salmonella, Campylobacter and, to a lesser extent, Yersinia, enteropathogenic Escherichia coli, Shigella and Vibrio parahaemolyticus; (ii) sporing rods, primarily Bacillus cereus and Clostridium perfringens. In approximately 90% of the outbreaks, foods of animal origin are the aetiological agent. The underlying faults in food handling are also the same as generally recognized and summarized in Table 1. Experts have termed these the "dual failure". It includes (i) contamination of the food from an animal or human source; (ii) allowing subsequent proliferation of the contaminants to the extent that the minimal infectious dose for the pathogen involved is exceeded. The aetiology of these

Table 1. Factors contributing to the pathogenesis of diseases of microbial aetiology transmitted by foods (%).[1] After F.L. Bryan (ref. 4).

Inadequate refrigeration	48
Storage at ambient temperature instead of refrigeration	34
Inadequate heat treatment (time-temperature-integral)	27
Contamination during food handling	23
Inadequate culinary re-heating	20
Improper storage at elevated temperature	19
Cross-contamination of raw foods	15

1. The sum of the percentages is more than 100 because as a rule more than one error, in fact mostly two - hence "dual failure" - are made during food preparation.

diseases is consequently so well known that control is
obviously within reach.

One striking fact, however, is that neither control, nor
even substantial reduction of morbidity has yet been achieved.
Hence, one might wonder whether these diseases are so trifling
that their prevention is barely worth much attention. This,
unfortunately, is totally untrue. The economic impact of
food-transmitted diseases is illustrated by the data in
Table 2. It is indeed tremendous and justifies intensive
preventive efforts. Moreover, there is often a considerable
loss of reputation for certain members of the food or catering
trade. As shown by the figures in Table 3, medically speaking,
these diseases are certainly not insignificant. Although
mortality is generally low and complications are relatively
rare, they are often rather serious when they occur.

Enteric pathogens spread by raw foods of animal origin
are the main culprits. Therefore, the much needed efforts
in prevention should concentrate on these commodities.
More specifically, the so-called <u>epidemiological</u>
<u>pressure</u> exerted by enteric pathogens of animal origin on food
and catering lines should be substantially reduced.

Table 2. Economic impact of microbial diseases transmitted by foods, compared to some leading causes of
human disease in general (in 10^9 US dollars per annum).

Malignant diseases	23
Trafic accidents	14
Coronary heart disease	14
Cardio-vascular accidents	6
Food-transmitted infections and intoxications[1]	1-10

1. Estimates arrived at by analysis of different basic
data by qualified epidemiologists.

6

Table 3. Complications observed after bacterial and parasitic enteric infections transmitted by foods.

Salmonellosis	cholecystitis, colitis, endocarditis, meningitis, myocarditis, rheumatoid syndromes, Reiter's disease, splenic abscesses
Yersiniosis	arthritis, erythema nodosum, spondylitis, septicaemia
Shigellosis	haemolytic-uraemic syndrome, synovitis
Campylobacteriosis	cholecystitis, endocarditis, meningism
Vibrio parahaemolyticus enteritis	septicaemia
Giardiasis	dystrophy, lymphoidal hyperplasia
Taeniasis	arthritis

3. ESSENTIALS AND PROSPECTS OF PREVENTION

3.1. The example set by the dairy and egg industries

The dairy industry has completely mastered the problem expounded at the end of the previous section. Since the early twenties, pasteurization of raw milk has been systematically introduced to protect the consumer against milk-borne infectious diseases (42). Subsequently, liquid milk and likewise dairy products manufactured from pasteurized raw milk have shown a conspicuously good health record (40). Unfortunately, a still prevailing, but medically quite unjustified opposition to mandatory pasteurization of raw milk is a threat to this situation. It would be very wise indeed if the dairy industry and public health officials alike would reject this pressure. Where, for technical reasons, the use of raw milk seems to be required, as in the manufacture of certain types of cheese, bactofugation instead of a heat treatment should be used to control bacterial pathogens (27).

In the egg products industry exactly the same phenomenon occurred. The introduction of legislation requiring pasteurization of egg products resulted in a dramatic decline in the importance of these products as vehicles in human salmonellosis (16).

7

3.2. The victimization of the meat and poultry industry

For meat and poultry a somewhat similar preventive approach has been recommended since about 1950. More particularly, the unchecked importation of extracted commodities such as fish-meal, meat-and-bone meal and cotton seed flour from areas with endemic enteric disease was strongly discouraged by the medical and veterinary professions. Mandatory heat treatment of such products upon entry to the country of import was emphatically recommended but, except in Denmark, never adopted. The uncontrolled introduction of pathogens with such feed components resulted in the predicted marked increase in the epidemiological pressure on the environment. This, in turn, led to the majority of slaughtered animals carrying these pathogens in their in-

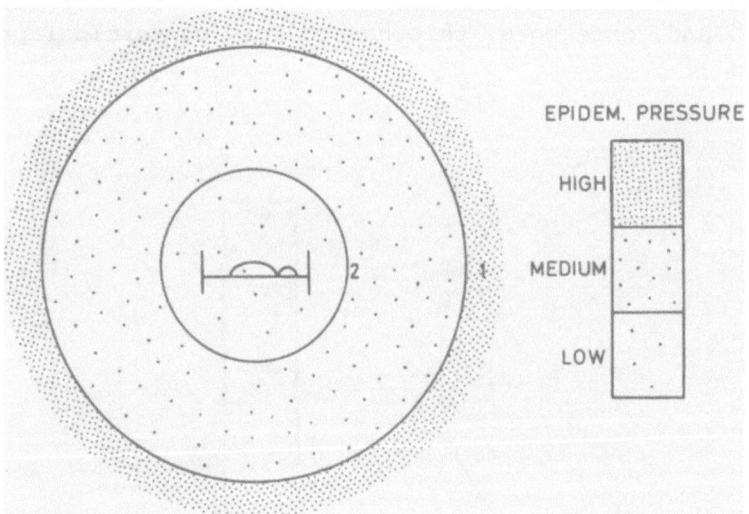

FIG. 1. The present situation with regard to the bacteriological safety, particularly of fresh pork and poultry.
The lines of defence representing optimal husbandry practices and hygiene in the slaughter-house (indicated as 1 in the diagram), and hygienic practices during culinary preparation (defence line 2) are not sufficient to guarantee pathogen-free commodities at the table. Consequently disease is caused when they are eaten.

8

testinal system. Because such carriers are not clinically ill,
they pass veterinary inspection. This clearly results in a
very heavy epidemiological pressure on abattoirs and slaughter
lines to the extent that it cannot be reduced to safe levels even
by the most sophisticated measures of hygiene so far developed
and practised (37, 12).

Quite obviously, because the first line of defence against
enteric pathogens contained within the slaughter-house is inadequate, the food sold to the housewife is dangerously contaminated. In spite of all the efforts made to educate her in
this field, this has neither been entirely successful in preventing the enteric pathogens of animal origin from reaching the ulti-
mate consumer (Fig. 1). Ecological studies have quite explicitly
demonstrated that the outlook for correcting the primary blunder
is quite dim (11, 36) - the infection cycles in the environment
have become autonomous (7, 8, 32, 15). Consequently, as in the
dairy and egg industries a third line of defence is required
(cf. Fig. 2). And, once more, this has to rely on terminal proces-

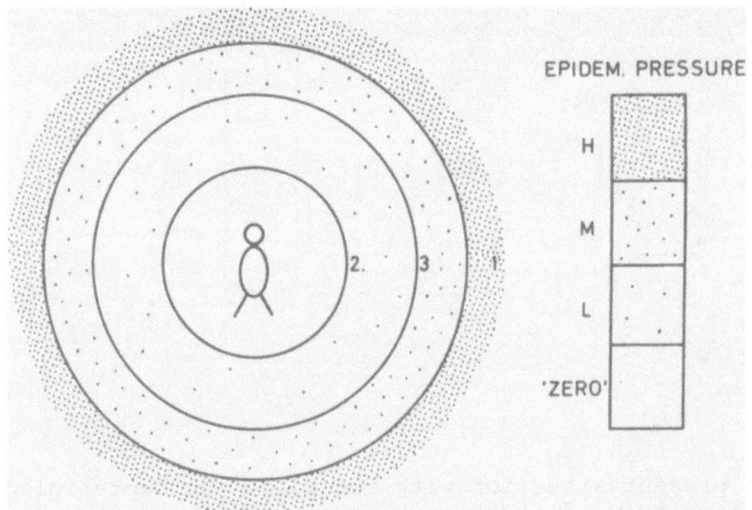

FIG. 2. The ideal situation: an almost completely pathogen-free
product is marketed via terminal processing for safety as with
pasteurized milk.
The three lines of defence together now ensure safe products
reaching the table, and no disease is caused by their consumption.

sing for safety. This includes:

1. Adequate bactericidal treatment leading to the <u>elimination</u> of pathogens initially present;

2. Absence of <u>recontamination</u> which would nullify the effect of the treatment referred to under 1.;

3. Subsequent storage under conditions which prevent <u>proliferation</u> of the very low numbers of pathogenic organisms surviving processing for safety.

3.3. <u>Better late than never</u>

Various processes have been developed for the processing of raw meat and poultry, leading to a high degree of elimination of enteric pathogens.

Heat treatments studied include the use of hot water, steam (10) and surface processing with infra-red radiation (37). These processes can all be used to achieve a sufficiently elevated degree of elimination of pathogens. However, these treatments suffer from two drawbacks: (i) any heat treatment changes the organoleptic characteristics of fresh food of animal origin; (ii) processing by heat requires large amounts of energy.

Various disinfectants acceptable to the food industry, such as chlorine, have also been used in attempts to reduce the numbers of pathogens on raw foods of animal origin. These have been found to be rather limited in application. Their bactericidal effects are often disappointingly low; some pose problems of physiological acceptance; while others, despite the demonstrated absence of toxicity, are nevertheless forbidden by legislation.

Thus, the use of ionizing radiation for this purpose has been studied very profoundly since about 1950. Its usefulness for the elimination of non-sporing enteric pathogens from meat and poultry surfaces will be assessed in the next section. A suggested term for this method of processing is radicidation (13).

4. RADICIDATION - NEITHER PANACEA NOR PESTILENCE

4.1. Bactericidal effects

Independently, Minck in 1896 (20) and Prescott in 1902 (34) discovered the bactericidal effect of ionizing radiation. It has now been well established that a dose of between 3 to 5 kGy (0.3 and 0.5 Mrad) results in a reduction of Enterobacteriaceae colony-forming units in the order of magnitude of 5 to 7 (17, 19, 26). When evaluated by the modern approach, termed risk analysis, this represents an adequate amount of reduction of epidemiological pressure (26, 27).

Such a dose of ionizing radiation does not result in a significant loss of organoleptic properties of raw meats or

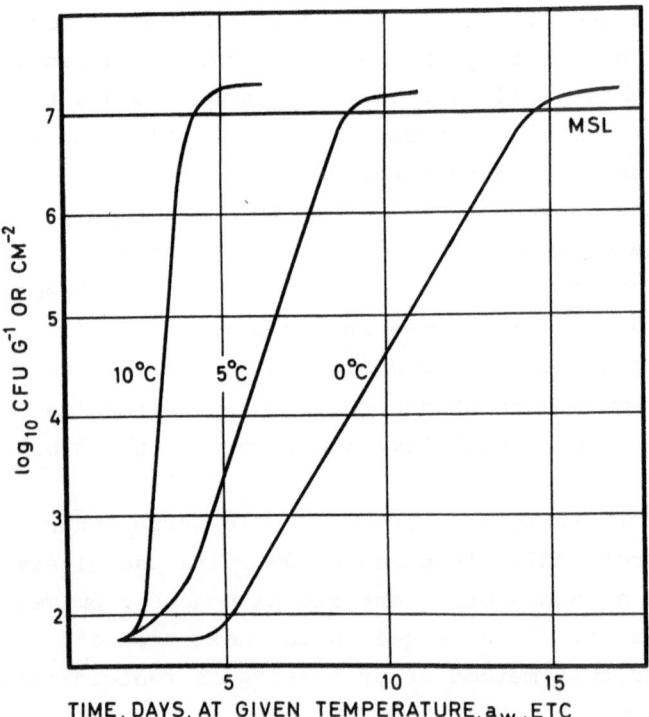

FIG. 3. The effect of temperature of chilled storage on time to spoilage, i.e. days to reach the minimum spoilage level (ca. 10^7 dfu cm^{-2}-MSL).

poultry. Likewise, radicidation at this level does not lead
to radiolytic changes that result in the formation of toxic
compounds. Investigations in this area have included assaying
for the production of mutagenic and teratogenic compounds;
these results were also totally negative (1, 9). Finally, it
is very important to note that radicidation can be used in
such a way that no induced radioactivity will result from the
treatment.

The processing of raw food of animal origin by radicidation

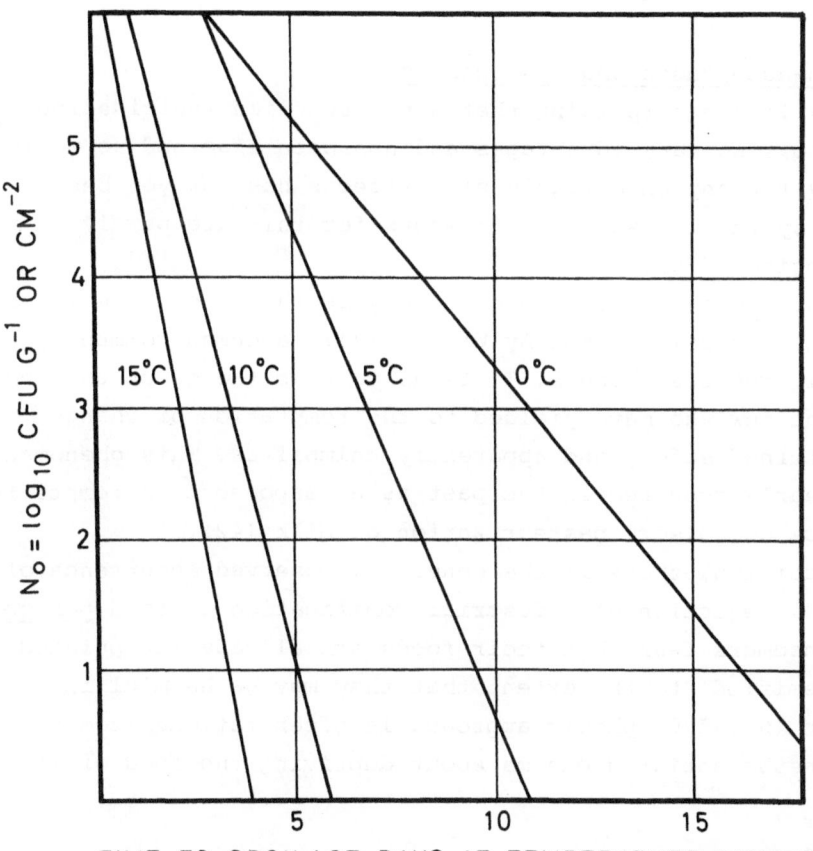

TIME TO SPOILAGE, DAYS, AT TEMPERATURE INDICATED

FIG. 4. The effect of initial load of psychrotrophic Gram-
negative rod-shaped bacteria (N_O) on time to attain the
MSL at different temperatures of chilled storage.

has two additional advantages. First and foremost it can be
applied to packaged units, precluding post-process recontamination which is always a threat to any processing for safety
(Section 3.2.). Furthermore, irradiation at a dose of 3 to
5 kGy leads to a reduction of colony-forming units of the
psychrotrophic Gram-negative rod-shaped bacteria that limit
the keeping quality of fresh meats and poultry under refrigeration in the order of magnitude of ca. 10^5 (30).
This results in the extra benefit of an increase in the keeping quality of these commodities by a factor of 3 to 5, as
demonstrated in Fig. 3 and 4[1].

4.2. Non-acceptance and its reasons

It is indeed surprising that a process like radicidation
which shows so many advantages and has been examined so meti-
culously for any undesirable side-effects has not yet been
adopted by the industry. The reasons for this are purely
psychological (25).

One of the mechanisms involved may be the general resistance
to change. As pointed out by Wilson (41), a human community
generally resists changes. It tends to keep the group unchanged
until the few who have yielded to the temptation of change
have returned safely and apparently unimpaired. This phenomenon
also clearly occurred in the past as a response to attempts to
introduce compulsory pasteurization of milk (42).

In addition, there is the generally observed occurrence of
emotional rejection of industrial modification of food per se.
Many consumers fear that their foods are already manipulated
and 'chemified' to the extent that they may be harmful in the
long run (6). This phobic approach is often rationalized by
fully unsubstantiated claims about depriving the food of its

---

1. The numbers of colony forming units (cfu) are expressed
per g or per cm^2, dependent upon the food, which can be solid
or may concern a surface area (e.g. of meat carcasses).

natural nutritive value, or the addition or formation of
hazardous substances. This was also displayed by sectors of
the public when milk pasteurization was introduced (40). It
must be admitted that a recent, comprehensive survey on the
aetiology of cancer does indeed indicate that food may play
a significant role (5). And incidents like the massive outbreak of life-threatening pneumonopathy after the consumption of allegedly falsified olive oil in Spain (39) do little
to reassure the alarmed consumer, or reinforce his feelings
of security that public health authorities are sufficiently
alert ...

Last but certainly not least there is the phobic rejection
of nuclear energy grosso modo. This is occurring on such a
large scale that we must assume that mankind was conditioned
by the traumatic incident of the first use of nuclear energy
for destructive purposes, but is also more inherently programmed
to phobias directed at physical phenomena of vital importance
to the survival of Homo sapiens (35). If this is true, then
any direct attempts to reduce radiophobia are unlikely to be
successful. Lacking training and experience in medical psychology,
the present authors can hardly be expected to analyse these
attitudes towards irradiated foods - let alone indicate how
to restore the consumers' confidence. But it would be utterly
unwise to dismiss the subject as a trifling one, or to deny
that the Public Health bacteriologist can do anything about it.

4.3. The key to the solution

The first step in guiding consumers and authorities alike
towards acceptance of radicidation as a valuable contribution
to processing of food for safety is to provide sufficient
reassurance with regard to the absence of detrimental effects
of foods thus treated. The recent recommendations of a WHO/
FAO Panel may serve as a basis for this approach (44). In this
context it should also and always be stressed that radicidation
will never be allowed to be used to cover up deficient manufacturing practices. In other words, exactly like pasteurization
in the dairy and egg industry, radicidation should become a

legitimate part of measures aiming at protection of the consumer
along the food line.

Next it is the ethical duty of the Public Health bacteriologist
to make alternatives to radicidation available. There are at
least two reasons for this obligation. First of all it acknow-
ledges scientists' acceptance of the consumers' doubts about
safety of foods, or of procedures applied to foods, even the
irrational ones. Moreover, it demonstrates that science is not
obsessively pressing for one particular mode of processing
food for safety. The only thing science is attempting is to
provide safe food for the consumer, and not at any price.

The latter considerations prompted us to study whether
a treatment of meat and poultry carcasses and portions with
non-toxic decontaminating agents would be a promising alternative to radicidation. Lactic acid was chosen for this purpose (22, 31, 28, 33, 38). It is regularly consumed in large
quantities with foods generally considered as wholesome, such
as cheese, yoghurt, buttermilk, sauerkraut and salami types
of sausage. The results obtained so far (29) are summarized
in Tables 4 and 5. They illustrate that a short treatment
of meat surfaces with 0.2 M lactic acid results in a reduction of enteric pathogens in the order of magnitude of
2 to >5. Further investigations have shown that any shift in
microbial community structure secondary to lactic acid treat-

Table 4. The lethal effect on pure cultures of exposure to
0.2 mol/litre lactic acid at pH ca. 2.5 for 2 minutes.

Class of pathogen	Number of strains tested	$\Lambda = \log_{10} N_o - \log_{10} N_f$
Salmonella	4	4.3 - 5.2
Yersinia	2	3.8 - 5.1
Escherichia coli	4	2.2 - 4.1
Campylobacter	2	> 5.3

Table 5. Flora shift of the psychrotrophic association of fresh and stored (5 days at 7 °C) pork skin, after (i) inoculation with porcine colon contents; and (ii) decontamination, as specified in Table 4.

Substrate	Lethality[1] (Λ)		σ_{ps} = flora shifts, psychrotrophs		
	colony-forming units (APT[2], 10 d, 7 °C)	Entero-bacteriaceae (37 °C)	G +ve[3] / G -ve ($\log P_f - \log P_o$)	G +ve Cat +ve[4] (%)	Cat -ve
Fresh pork skin	1.4	2.2	>0.6	− 64	+ 64
Stored pork skin	2.1	2.8	>0.9	− 25	+ 25

1. As in Table 4, $\Lambda = \log_{10} N_o - \log_{10} N_f$
2. APT = all purpose tween agar
3. G +ve = Gram-positive; G -ve = Gram-negative
4. Cat = catalase

ment was toward an association of lactic acid bacteria. This substantiates previously published findings (43). It is a most fortunate side-effect of lactic acid decontamination treatment that it results in a protracted protection of meat thus treated against colonization by organisms of health significance. In this respect, lactic acid decontamination is somewhat similar to radicidation where extended protection results from packaging before processing for safety.

5. EPICRISIS

Even the most advanced countries of the world suffer from a tremendous incidence of food infections attributable to food of animal origin. The underlying mechanism of this morbidity is known and it is, therefore, a pressing assignment to design measures to reduce the incidence.

The classical defence lines relying on high-level hygiene in the slaughter-house and during food preparation are clearly insufficient in controlling these food-transmitted enteric infections. Consequently, a third line of defence, processing for safety, is required here as it was in the dairy and later the egg products industry. At least two effective modes of processing for safety of raw food of animal origin are available to erect this third line of defence. These include radicidation and lactic acid decontamination. The two procedures lead to a marked reward in terms of control of infection, while the consumer can be reassured that they are both fully safe.

It is hence the duty of the Public Health bacteriologist to advise regulatory agencies and consumers in chosing one, or both, of these measures for improved health protection. Obviously, the bacteriologist cannot possibly hope to achieve this goal without recruiting the assistance of experts in human behaviour and more particularly in factors determining selection and rejection of foods by consumers.

6. REFERENCES

1. Barna, J., Compilation of bioassay data on wholesomeness of irradiated food items. Acta alimentaria 8 (1979) 205-315.
2. Beckers, H.J., Foodborne infections and intoxications in the Netherlands, 1977-1979. Nederl. Tijdschr. Geneesk. 125 (1981) 2167-2170.
3. Bouwer-Hertzberger, S.A. & Mossel, D.A.A., A study of the morbidity and aetiology of diseases transmitted by foods in the region of Haarlem. Nederl. Tijdschr. Geneesk. 124 (1980) 1460-1463.
4. Bryan, F.L., Foodborne diseases in the United States associated with meat and poultry. J. Food Protection 43 (1980) 140-150.
5. Doll, R. & Peto, R., The causes of cancer: quantitive estimates of avoidable risks of cancer in the United States today. J. Natn. Cancer Inst. 66 (1981) 1191-1308.
6. Douglas, M.T., Purity and danger. An analysis of concepts of pollution and taboo. London, Routledge & Kegan Paul, 1978.
7. Edel, W., Guinee, P.A.M., Schothorst, M. van & Kampelmacher, E.H., Salmonella cycles in foods with special reference to the effects of environmental factors, including feeds. J. Canad. Inst. Food Sci. Technol. 6 (1973) 64-67.
8. Edel, W., Schothorst, M. van & Kampelmacher, E.H., Epidemiological studies on Salmonella in a certain area ("Walcheren Project"). I. The presence of Salmonella in man, pigs insects, seagulls and in foods and effluents. Zentralbl. Bakt. Parasitenkd. Abt. I, Orig., A, 235 (1976) 476-484.
9. Elias, P.S., The wholesomeness of irradiated food. Ecotoxicol. environm. Safety 4 (1980) 172-183.
10. Eustace, I.J., Control of bacterial contamination of meat during processing. Food Technol. Aust. 33 (1981) 28-32.
11. Foster, E.M., The need for Science in food safety. Food Technol. 26 (1972) 81-87.
12. Gerats, G.E., Bacterial contamination of pig carcasses during the slaughter-process with special reference to the impact of severing the intestines. Presented at the Winter Meeting of the Society of Applied Bacteriology, (London, 13 January 1982) J. appl. Bact. (in press).
13. Goresline, H.E., Ingram, M., Macuch, P. Mocquot, G., Mossel, D.A.A., Niven, C.F. & Thatcher, F.S., Tentative classification of food irradiation processes with microbiological objectives. Nature 204 (1964) 237-238.
14. Huisman, J., Microbiële voedselvergiftiging en voedsel-infectie. Epidemiologie en preventie. Alphen a.d. Rijn, Stafleu, 1980, pp. 14-15.
15. Kampelmacher, E.H., Public health aspects of food irradiation. In: Food Irradiation Now, The Hague, Nijhoff (1982) 20-39.
16. Lee, J.A., Recent trends in human salmonellosis in England and Wales: the epidemiology of prevalent sero-types other than Salmonella typhimurium. J. Hygiene 72 (1974) 185-195.

17. Ley, F.J., Kennedy, T.S., Kawashima, K., Roberts, D. & Hobbs, B.C., The use of gamma radiation for the elimination of Salmonella from frozen meat. J. Hygiene 68 (1970) 293-311.
18. Mann, J.M., A prospective study of response error in food history questionnaires: implications for foodborne outbreak investigation. Am. J. Public Health 71 (1981) 1362-1366.
19. Maxey, R.B. & Tiwari, N.P., Irradiation of meats for public health protection. In: Radiation preservation of food. International Atomic Energy Agency, Vienna, Publ. STI/317 (1973) 491-503.
20. Minck, F., Zur Frage der Wirksamkeit der Röntgenstrahlung aug Bakterien sowie die Möglichkeit ihrer eventuellen Anwendung. Münch. medizin. Wschr. 5 (1896) 101.
21. Mossel, D.A.A., The destruction of Salmonella bacteria in refrigerated liquid whole egg with gamma radiation. Internat. J. appl. Radiation Isotopes 9 (1960) 109-112.
22. Mossel, D.A.A. & Bruin, A.S. de, The survival of Enterobacteriaceae in acid liquid foods stored at different temperatures. Annls. Inst. Pasteur Lille 11 (1960) 65-72.
23. Mossel, D.A.A., Büchli, K. & Waart, J. de, Doserange-finding tests for the elimination of salmonellae from proteinaceous foods with a low a_w by irradiation with Co-60 gammarays. Antonie van Leeuwenhoek 31 (1966) 31-220.
24. Mossel, D.A.A., Krol, B. & Moerman, P.C., Bacteriological and quality perspectives of Salmonella radicidation of frozen boneless meats. Alimenta 11 (1972) 51-59.
25. Mossel, D.A.A., Health protection aspects of food irradiation at the pasteurization level. Acta Alimentaria, 6 (1977) 253-262.
26. Mossel, D.A.A., The elimination of enteric bacterial pathogens from food and feed of animal origin by gamma irradiation with particular reference to Salmonella radicidation. J. Food Qual. 1 (1977) 85-104.
27. Mossel, D.A.A., Microbiology of foods. The ecological essentials of assurance and assessment of safety and quality, third edition, Utrecht University Press, 1982.
28. Mulder, S.J. & Krol, B., Der Einfluss der Milchsäure auf die Keimflora und die Farbe frischen Fleisches. Fleischwirtschaft 55 (1975) 1255-1258.
29. Netten, P. van & Mossel, D.A.A., The ecological consequences of decontaminating raw meat surfaces with lactic acid. Arch. Lebensm. Hyg. 31 (1980) 190-191.
30. Niemand, J.G., Linde, H.J. van der & Holzapfel, W.H., Radurization of prime beef cuts. J. Food Protection 44 (1981) 677-681.
31. Ockerman, H.W., Borton, R.J., Cahill, V.R., Parrett, N.A. & Hoffman, H.D., Use of acetic and lactic acid to control the quantity of micro-organisms on lamb carcasses. J. Milk Food Technol. 37 (1974) 203-204.
32. Oosterom, J., Erne, E.H.W. van & Schothorst, M. van, Studies on the possibility of fattening pigs free from Salmonella. Neth. J. vet. Sci. 106 (1981) 599-612.

33. Patterson, J.T. & Gibbs, P.A., Vacuum-packaging of bovine edible offal. Meat Science 3 (1979) 209-222.
34. Prescott, S.C., The effect of radium rays on the colon bacillus, the diphteria bacillus and yeast. Science 20 (1904) 246-248.
35. Seligman, M.E.P., Phobias and preparedness. Behav. Ther. 2 (1971) 307-320.
36. Silliker, J.H., Status of Salmonella - ten years later. J. Food Protection 43 (1980) 307-313.
37. Snijders, J.M.A. & Gerats, G.E., Hygiene bei der Schlachtung von Schweinen. VI. Die Verwendung eines Infrarottunnels in der Schlachtstrasse. Fleischwirtschaft 57 (1977) 2216-2219.
38. Snijders, J.M.A., Schoenmakers, M.J.G., Gerats, G.E. & Pijper, F.W. de, Dekontamination schlachtwarmer Rinder- körper mit organischen Säuren. Fleischwirtschaft 59 (1979) 656-663.
39. Tabuenca, J.M., Toxic-allergic syndrome caused by in- gestion of rapeseed oil denatured with aniline. Lancet II (1981) 567-568.
40. Westhoff, D.C., Heating milk for microbial destruction: a historical outline and update. J. Food Protection 41 (1978) 122-130.
41. Wilson, G.D. (Ed.), The psychology of conservatism. London, Academic Press, 1973.
42. Wilson, G.S., The necessity for a safe milk-supply. Lancet II (1933) 829-832.
43. Woolford, M.K., Microbiological screening of food pre- servatives cold sterilants and specific antimicrobial agents as potential silage additives. J. Sci. Food Agric. 26 (1975) 229-237.
44. World Health Organization, Wholesomeness of irradiated food. Tech. Rep. Ser. WHO No 659, 1981.

Public health aspects of food irradiation

E.H. Kampelmacher

1. INTRODUCTION

When debating the public health aspects of food irradiation
it is necessary in the first place to define which aspects are
being discussed due to the fact that these are two-fold:
toxicological and radiological. Equally, the effect of food
irradiation as a convenience to the consumer also needs to be
discussed, i.e. the prevention of food deterioration and also
the prevention of disease that could be passed on to the consumer by ingestion.

On the other hand, the effects that could possibly be created by the application of radiation and which need to be
evaluated by toxicological and microbiological research must
also be included in the public health aspects.

It will be obvious that I cannot elucidate exhaustively on
such a broad spectrum of effects in the time allotted to me.
I shall, therefore, limit myself to a few essentials, working
on the theory that for all the various topics to be discussed
a liberal choice of literature is available.

2. DETERIORATION OF FOODSTUFFS

The fight to prevent food from deteriorating is probably
as old as mankind itself. Since pre-historic times man has,
in times of plenty, tried to save his food for times of scar-
city. In spite of the various methods used in trying to prevent
deterioration, the FAO has estimated that a third to a quarter
of the world's food production is lost due to this problem.
This is particularly poignant at a time when seventy per cent
of mankind is starving or near-starving. Whilst methods such

Food Irradiation Now – ISBN 90-247-2703-0
Copyright Gammaster and Martinus Nijhoff/Dr W. Junk Publishers

as freezing, cooling and heat preservation can be very effective
in preventing deterioration, the use of energy is also important.
Energy, technical installation and know-how in the sub-tropical
and tropical producing countries, where the need is many times
greater than in our temperature climate, is only available to
a very limited degree, if at all. Considering the world-wide
energy shortage, a solution to the deterioration problem by
using the methods just mentioned on a large scale, in areas
where prevention of deterioration is most needed, cannot be
foreseen in the near future. It would be incorrect to state
that food irradiation is the solution to this problem, but this
technique could make an important contribution to the battle
and recently, in fact, has been successfully applied experimentally.

For places with a warm climate, this technique would mainly
be used for fruit and vegetables; products that, if only they
could be exported, could be economically significant for most
of the non-industrial countries. The treatment of grains,
which are not only threatened by microorganisms but also by
infestation, could also be important. In more temperate
climates, ionizing radiation has already been being used with
success for some time to inhibit the sprouting of potatoes,
also a form of deterioration. The advantage here is that the
alternative treatment with suspect chemicals can be avoided.
Applying low doses of gamma radiation to soft fruits in these
moderate zones could slow down their deterioration, thereby
improving the national and international marketing of these
products. Recent tests with, for example, potatoes and mushrooms have given very satisfactory results. Briefly, it can
be said that food irradiation has already proved to be a very
effective method of solving this urgent problem of food deterioration and, in many cases, can replace chemical treatments.
Further development concerning the possibilities of this application is on our doorstep and could be an effective tool in
combating the enormous losses caused by the deterioration of
meat, fish, fruit, vegetables and grains.

3. INFECTIONS IN FOOD AND FOOD POISONING

Even though infectious diseases in general, when compared with cancer or cardio-vascular diseases, are less significant now than a century ago, there are still two human activities, sexuality and food consumption, which are causing an alarming increase in infectious diseases throughout the whole world (Table 1) (1, 2). This trend is likely to continue for the present, since these diseases form part of a vicious circle, in which unemployment, poverty and malnutrition play a leading role in many parts of the world (Fig. 1) (3).

Many complex factors have been responsible for the increase in food poisonings (by this is briefly understood: infections in food caused by the microorganism itself and food poisonings caused by toxins of microorganisms) especially since the 2nd World War. Some of the factors responsible are the mass-breeding and mass-fattening of animals, mass production and processing

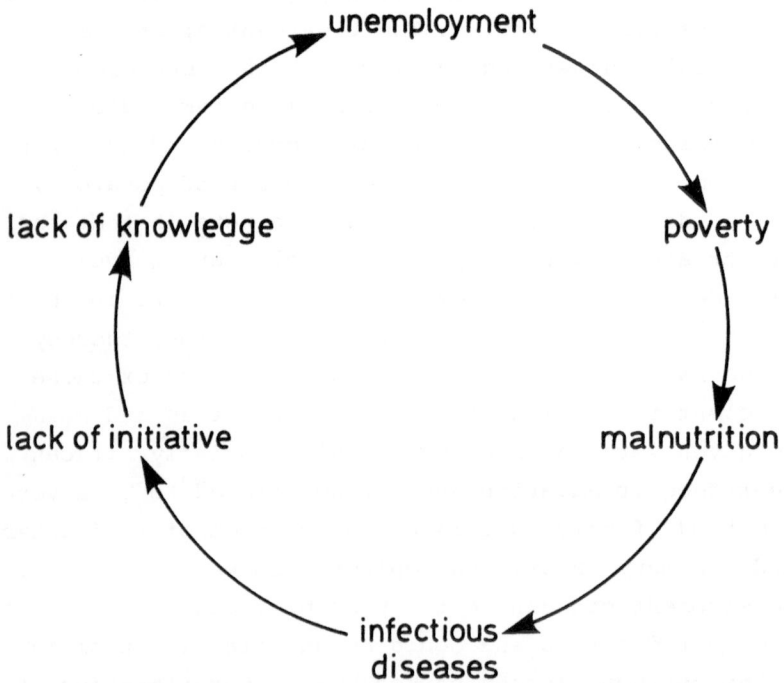

FIG. 1. The vicious circle (after Hedén, ref. 3).

Table 1. Incidence of infectious diseases in the Netherlands, 1978.

Gonorrhoea	9613 ⎱	10 614
Syphilis	1001 ⎰	
Salmonellosis		6 197
Parotitis epid.		1 105
Hepatitis epid.		1 041
Rubella		854

of foods, especially those of animal origin, the migration of millions of people (tourists, immigrant-labourers), the ever-increasing market in foodstuffs and animal feed, changing eating habits and, last but not least, the increasing environmental pollution which, amongst other things, results in the contamination of human foodstuffs and animal feed.

By far the most important form of food poisoning in most countries of the world is salmonellosis. In spite of the hygiene regulations in slaughter-houses, factories and kitchens and in spite of the tight control of foodstuffs and animal feed, the number of cases of Salmonella in man either remain permanently high or are still rising. In the Netherlands, an average of approximately 7000 cases of salmonellosis in humans are registered annually (Figs. 2 and 3). According to cautious estimates, this is only 1 - 5% of the actual morbidity, which means that there are a further few hundred thousand cases even though, on average, they may be of a less serious nature than the officially registered ones, and are often simply diagnosed as either a summer influenza or an abdominal influenza (4). We see the same picture in the USA where approximately 25 000 cases are registered annually, but the real number of cases is estimated to be two million. Apart from human suffering, sometimes ending in death, this morbidity also results in economic losses due to hospitalization, medical costs and lost working days. This amounts to at least $ 300 million (5, 6). The prevalence of Salmonella in man, animals, foodstuffs and

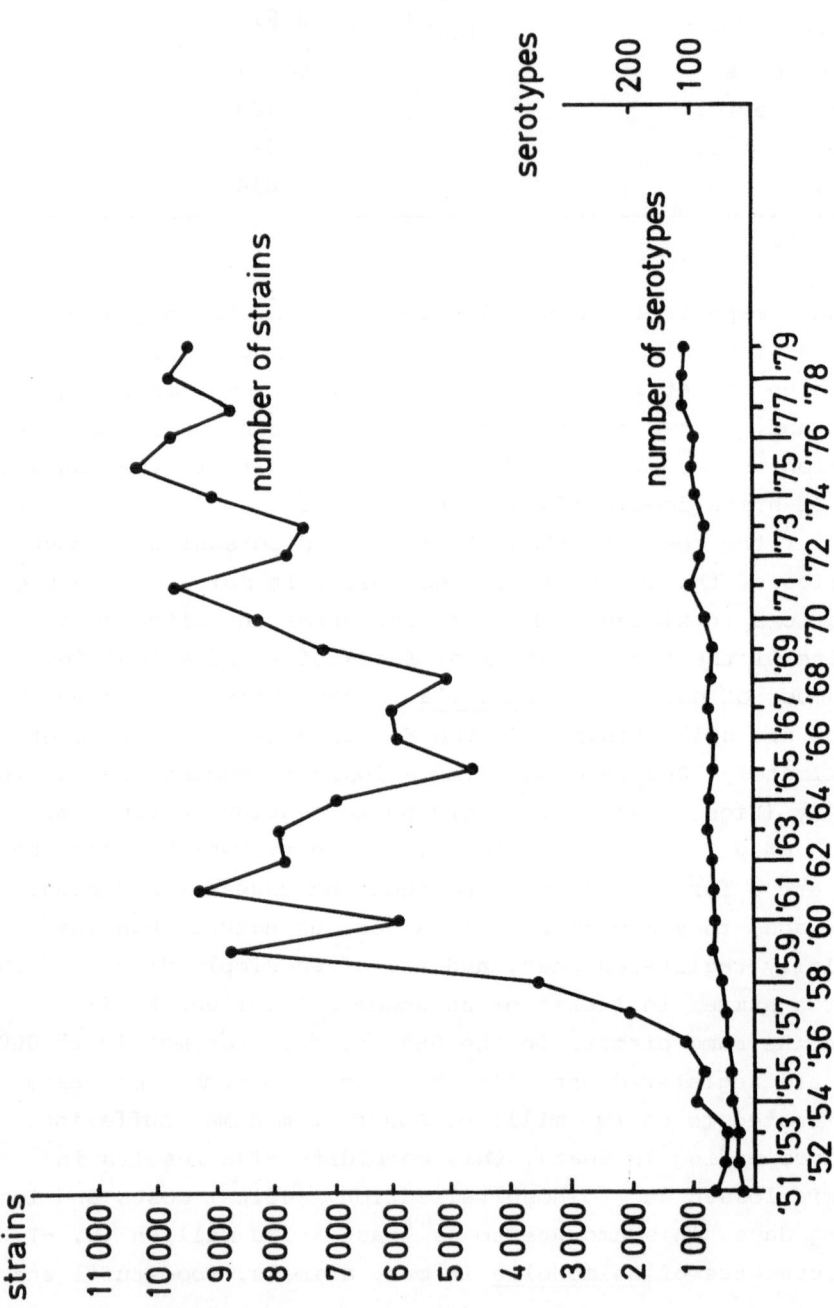

FIG. 2. Number of Salmonella strains and number of serotypes isolated from men in the Netherlands 1951-1979.

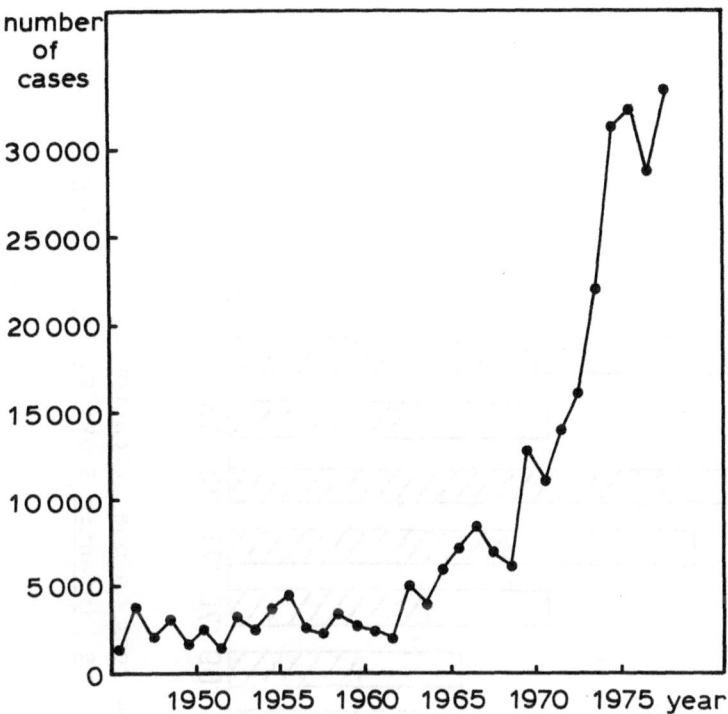

FIG. 3. Salmonellosis in humans, Federal Republic of Germany.

feed and the environment has, in the past twenty-five years,
been studied extensively in many parts of the world, and many
publications are available (7, 8, 9). Briefly, it can be stated
that it is primarily foodstuffs of animal origin, such as
pork, veal, poultry, game, shellfish, fish, egg products and
milk products, in areas where pasteurization is not common,
that may become contaminated with Salmonella (Fig. 4). But
various vegetable products such as grated coconut, spices,
vegetables and even processed products such as chocolate, can
contain Salmonella and lead to disease in man (14, 15, 16, 17,
18). These contaminations are part of a vicious circle of
infection, in which contaminated water and soil, birds, insects
and rodents, contaminated foodstuffs, infected animals (which
do not usually show any symptoms and, consequently, as so-called
clinical carriers are not recognized during routine inspections),

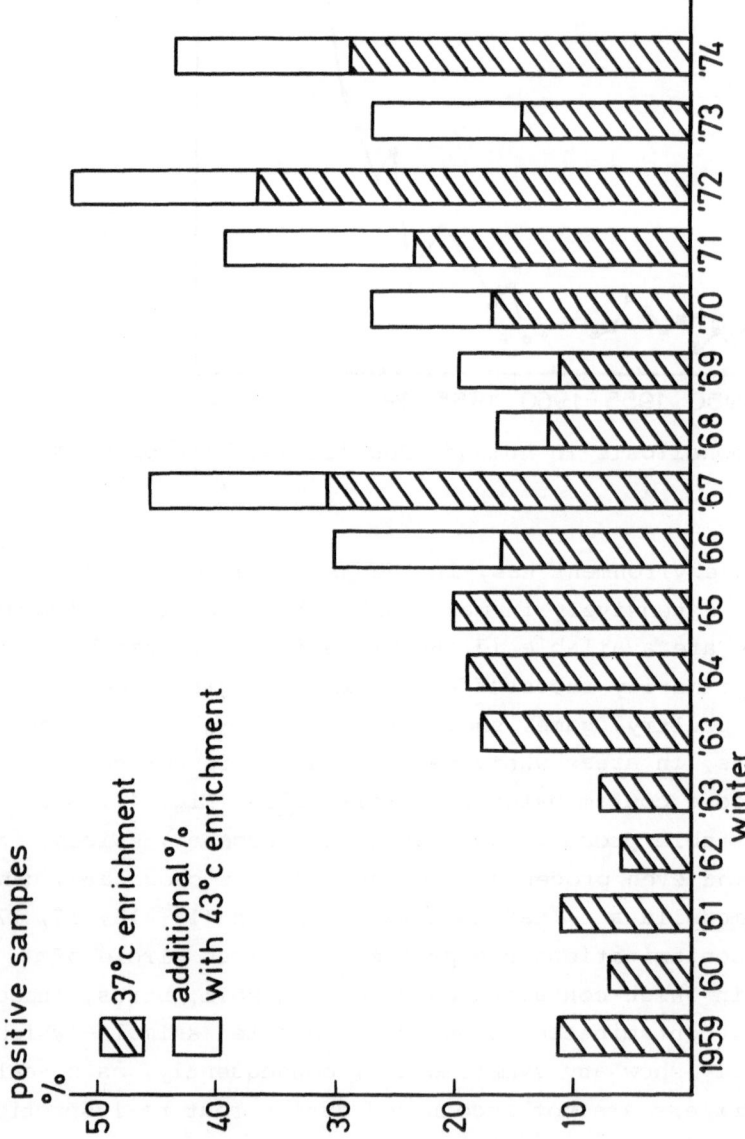

FIG. 4. Salmonella in minced meat (yearly random test in one Dutch town).

Serotypes: <u>S. typhimurium</u> 34.7%, <u>S. panama</u> 17.1%, <u>S. infantis</u> 5.9%, <u>S. livingstone</u> 4.6%, <u>S. stanley</u> 3.5%, other types (36) 34.2%.

contaminated foodstuffs, and infected humans who discharge
excrement into the environment, play a leading role (10, 11,
12, 13). Epidemiological studies have demonstrated that due
to the very high degree of contamination (some investigations
have shown, for example, that up to 40% of minced meat samples
and up to 70% of the examined poultry contained Salmonella),
secondary contamination during production, processing and
preparation in the kitchen plays a much greater role in human
infections than primarily contaminated products (4, 19, 20,
21, 22). This is especially true of products such as minced
meat, hamburgers and sausages, which either are consumed raw
or are often insufficiently heated. Frozen products such as
chickens, and the released drip-water can also play a major
role. Contaminated surfaces, kitchen utensils and, last but
not least, human hands can cause cross-contamination, whereby
Salmonella, at favourable temperatures, can increase within
a few hours to infectious doses of 10^5 to 10^7 bacteria (Gram)
(10, 23, 24) in products that have already been cooked, such
as chicken, meat and milk products and which have not been
cooled immediately. If this happens in large kitchens belonging
to hotels, hospitals, old peoples' homes, etc. then there is
a threat to large groups of consumers which is, in fact, what
happened recently at a meeting of the European Top in Maastricht,
the Netherlands (24a).

The above-mentioned facts concerning salmonellosis is also,
mutatis mutandis, true of campylobacteriosis, an infection
that, in the past few years, has become increasingly important
as far as food poisonings are concerned and which, epidemiologically, is almost the same as salmonellosis. Food of animal
origin also plays a leading role, together with secondary and
cross-contamination. It is becoming clear that campylobacteriosis,
due to increased interest in this disease and, therefore, more
tests on patients and foodstuffs, is going to take up a more
important position under the heading of food poisonings.

One disease that affects certain areas of the world, and
especially where there is mass-preparation of food, is staphylo-
enterococcosis, caused by the enterotoxinogenic types of

Staphylococcus aureus. These microorganisms, that are found in
the mucous membranes of the mouth and nose of healty individuals
and in lesions, especially on the fingers and inside the nails,
and also in animals and are, therefore, present in food of
animal origin, from a heat-stable toxin that can lead to a
very short but sometimes very serious illness. The multiplication
of staphylococci and the production of toxins occurs at temperatures above 18 ^{O}C, and the toxin, once formed, is not inactivated
by a normal heat treatment such as cooking.

Protein-rich food, such as meat and meat products and milk
products, stimulates the growth of staphylococci at favourable
temperatures and are, therefore, a potential source of serious
infections for large groups of consumers (25, 26, 27, 28).

There are also food poisonings caused by microorganisms of
the genus Clostridium, especially in the western hemisphere.
C. perfringens should also be mentioned here. This microorganism is ubiquitous and is frequently found in human and
animal faeces and is, therefore, also present in food of
animal origin, especially meat and meat products. Under favourable
conditions, the heat-resistant spores of this microorganism
can survive the cooking process and develop into large numbers
of vegetative cells. Large pieces of meat and large turkeys,
mashed potatoes and soups, when kept in sizeable containers with
insufficient oxygen, are excellent growth media for these
anaerobic bacteria (10, 28, 29). Primary as well as secondary
contamination of the meat during slaughtering and further
processing is possible. Certain cooking techniques, such as
the preparation of food without oxygen, for example meat pies,
and insufficient heating of left-overs, are also involved in
the occurrence of food poisoning by C. perfringens. A much
more serious, but fortunately less frequent form of food
poisoning is the one caused by C. botulinum. This extremely
serious infection which, without rapid and proper treatment,
leads to death, is caused by C. botulinum toxins formed in
food kept under strict anaerobic conditions, such as home-
bottled fruit and vegetables and vacuum-packed meat and fish
products. Type E, which is capable of producing toxins at

refrigerator temperature (ca 5 $^{\circ}$C) should be mentioned here.
Even though this production develops slowly compared with the
simultaneously developing psychrophilic spoilage organisms,
the potential danger of C. botulinum, type E, needs to be
stressed. Recent epidemiological research has shown that
C. botulinum, together with Salmonella, occurs in an environment in contamination cycles, which means that the already
existing threat of this serious form of poisoning must be
permanently taken into account (30, 31, 32, 33).

A type of food poisoning that occurs less frequently in
the West, but which has claimed victims in Japan and Southeast
Asia, is the one caused by Vibrio parahaemolyticus.
Contaminated fish and fish products, as well as other seafoods,
which have been either insufficiently cooked or eaten raw, are
responsible for outbreaks of infections in humans in these
regions. A further complication is the fact that V. parahaemolyticus is not even inhibited, let alone destroyed, by the
relatively high concentrations of salt with which the fish is
often treated. This form of contamination has grown in importance
in the past twenty years or so in the regions just mentioned,
where fish is the most important source of protein for the
rapidly growing population and where it is not usual to cook
it (34, 35, 36).

Finally, some mention should be made of parasitic disorders,
in which food of animal origin plays an important role. Bladderworm and tapeworm infections in both humans and animals, especially in Africa and South America, the latter being a continent
from which Europe imports a lot of meat, are of great importance.
Toxoplasmosis can lead to serious disorders and we cannot rule out
raw or insufficiently heated foodstuffs of animal origin as
being the source of infection. Trichinosis also occurs from time
to time in various European countries, making the consumption of
raw pork potentially dangerous. Parasitic disorders in fish,
such as herringworm, are also important and, especially in
South and Southeast Asia, are a threat to human health (38,
39).

4. PREVENTION OF FOOD CONTAMINATION AND FOOD POISONING

We can conclude from all this that the cooling and heating
of foodstuffs can clearly reduce both the chance of contamination
and the occurrence of food poisoning. Cooling from production
to consumption is not always possible, resulting in the multiplication of microorganisms. Heating is likewise not always
possible, because certain products are consumed either raw
or only partially cooked. Contamination and rapid growth of
microorganisms in the often still warm food is one of the
most important factors in the occurrence of food poisoning.
Both cooling and heating require a lot of energy and, therefore,
cannot be used everywhere and at all times.

Another method of prevention is the addition of chemicals
to foodstuffs, but there are many serious toxicological argu-
ments against it. On the basis of this data we need to consider
the possibilities offered by food irradiation.

Intensive research in many countries has shown that
radicidation, defined as a treatment with ionizing radiation
to approximately 10 kGy (1 Mrad), is sufficient to adequately
reduce the number of viable, non-spore-forming microorganisms
in food (40). So much attention was given to the possible
detrimental effects of irradiation on human health that it
has been more thoroughly investigated, both toxicologically
and microbiologically, than any other food treatment. It was
also shown that the formation of radiolytic products cannot
be considered as a health hazard, because most of these products
can also be found in food that has not been irradiated, as
a result of other processing procedures. Tests with animals
from successive generations, which had been fed with irradiated
food, have shown that there are no harmful effects. The induction
of radio-resistant microorganisms appears to be a phenomenon
that can only be induced under laboratory conditions (41). So
far, where good manufacturing practices have been applied in
the poultry preservation industry, radio-resistance, in contrast
to e.g. resistance to antibiotics, has not fiven rise to any
difficulties (42). As a result of the many investigations
that have been carried out for several years in various countries,

a group of experts, within the framework of a WHO consultants'
meeting, have recently concluded that the irradiation of food
to 10 kGy is not associated with any toxicological threat,
and that toxicological evaluation of foods treated in this way
is no longer necessary (43, 44, 45, 46).

What, apart from the desired killing of microorganisms, are
the specific advantages of food irradiation? The first thing
to be mentioned is the unique possibility of irradiating food
after packaging, thus eliminating the serious problem of
cross-contamination during food production and preparation.
Secondly, products that are normally eaten raw, such as beef
tartar, raw sausage, and oysters, can be made biologically
safe which, considering the dangers of raw products of animal
origin, is of great importance. Furthermore, the addition of
suspect chemicals becomes unnecessary. Thirdly, the low energy
costs of the process should be emphasized. In the United States
the irradiation of food has been calculated to be 0.5 to 1.0
cent per kilogram. The irradiation process, therefore, requires
much less energy when compared with processes such as heating
and cooling.

It would be incorrect and scientifically irresponsible if
the disadvantages of food irradiation were left unmentioned.
On the one hand, these concern changes in the food that also
occur with most of the conventional methods of preservation,
such as change in consistency, colour, and vitamin content,
and on the other hand, specific radiation changes, such as
a certain colour, flavour, and aroma. These last-mentioned
effects are strongly dependent on the composition of the
product (e.g. high protein or fat content) and occur at doses
above 2.5 kGy. These undesirable organoleptic effects can,
however, be largely prevented if the product to be irradiated
has first been cooled or, preferably, frozen, especially if
there is an absence of oxygen (47).

Which foodstuffs should be given priority for an irradiation
treatment, especially when considering the elimination of
disease-producing organisms? As has been repeatedly stated,
foodstuffs of animal origin play a major role in the occurrence

of food infections and food poisonings. Due to the fact that poultry and game is frequently contaminated with especially Salmonella and Camplyobacter but also with Staphylococcus aureus and C. perfringens, the irradiation of poultry and game could be a first priority. The raising of poultry free from pathogenic microorganisms will not be accomplished in the near future. Due to mass-slaughtering (tens of thousands of birds per hour), cross-contamination is unavoidable, resulting in large numbers of infected poultry reaching the kitchen. In-plant chlorination in the slaughter-houses reduces the contamination pressure, but is not permitted in some countries. Steaming or pre-cooking of poultry has been tried experimentally, but on a large scale would be extremely difficult, together with the fact that such a product would not always be acceptable to the consumer (48, 49).

The packaging of poultry after slaughter and irradiating with doses of 2 to 5 kGy is extremely effective in the elimination of Salmonella and Campylobacter (50, 51). Our own investigations have shown that in 100 g chicken skin or 1500 ml drip-water, <2 to 1400 salmonellae occur, and <100 (5-60) salmonellae were found in 90% of the birds investigated (52). A reduction in Salmonella of 3 log cycles was reached with the above doses. If the birds are frozen there is an additional reduction of approximately half a log cycle. In the Netherlands, Canada and the USSR clearance has been given for test-marketing irradiated poultry. The WHO has declared irradiation up to a maximum dosage of 7 kGy as "unconditionally safe for human consumption" (44). Because of the Salmonella contamination of poultry in the Netherlands and the potential danger of this foodstuff with regard to the occurrence of food poisoning, large-scale irradiation would be a very important preventive measure (53).

What concerns poultry also largely applies to meat. Red meat not only contains disease-producing microorganisms; para-sites can also occur (54, 55). As with poultry, contaminated meat can cause cross-contamination in the kitchen, especially from the contaminated drip-water from frozen meat. Moreover,

minced meat, hamburgers and fresh sausage are often strongly
contaminated with disease-producing organisms; radiation in
low doses could be effective in destroying these dangerous
organisms. Meat, such as tartar, is often consumed raw
in the Netherlands and in the surrounding countries (56).
Recent investigations have shown that such products can be
highly contaminated with pathogens (57). Irradiation could,
in this case, be the only practicable way of ensuring that
these raw products are safe for the consumer (Fig. 5).

Parasites in meat, which in our part of the world mainly
concerns bladderworm and tapeworm as well as toxoplasmosis,
are effectively eliminated by freezing. By a combined process
of freezing and irradiation with 0.5 to 1 kGy, which can destroy
the infection capacity of many parasites, the freezing time could
be shortened considerably resulting in a saving of energy (58).
In countries where there are no freezing facilities, the
exclusive use of irradiation could be of great importance in
the prevention of these parasitic disorders in man (59).

A lot of research is being done into the possible irradiation
of fish, fish products, shell-fish and other seafoods. These
foodstuffs decay even more rapidly than meat, and in subtropical
and tropical areas are very important as they are the main
source of protein. Morever, due to the increasing contamination
of coastal waters, they often contain disease-producing microorganisms such as Salmonella, Parahaemolyticus, C. botulinum
type E and a number of parasites. Irradiating these products
with doses between 0.75 and 2.5 kGy can delay the decaying
process and eliminate pathogens (60, 61, 62).

Combined processes such as salting, drying, smoking and
irradiating with low doses have prolonged storage life and
eliminated disease with no adverse organoleptic effects (63).

In this connection, the good results obtained in the
Netherlands regarding the irradiation of frog legs should also
be mentioned. By applying a dose of 4 kGy to this product,
which is often infected with Salmonella, the pathogen can be
eliminated without impairing the organoleptic qualities (64,
65, 66, 67).

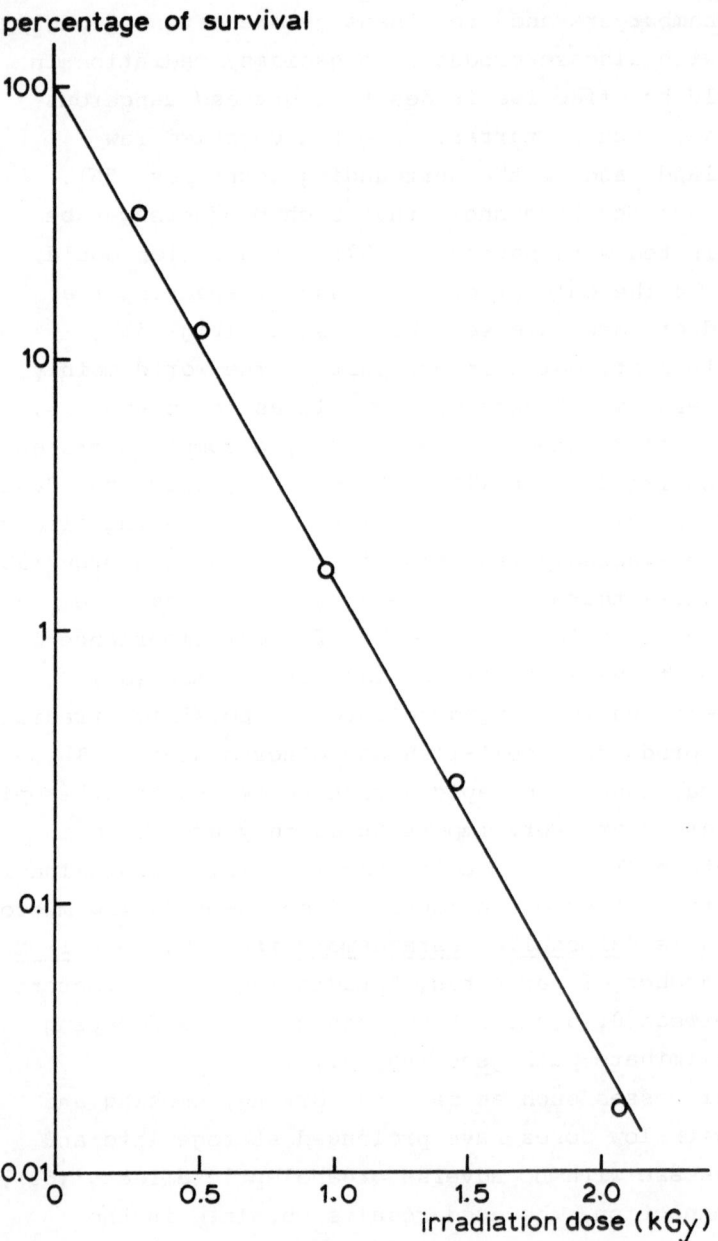

FIG. 5. Effect of irradiation on <u>Salmonella typhimurium</u> in ground beef; 7 D_{10} = 3.85 kGy (after Simard et al., ref. 55).

Commodities that do not have an animal origin but that would benefit from irradiation are spices. These products, that are used on a large scale in the meat industry, sometimes contain disease-producing bacteria and large numbers of decay-producing organisms that, under favourable conditions, can multiply in meat (68, 16). Decontamination is, therefore, essential and ethylene oxide is used on a large scale use of this gas, irradiation with doses of 5 to 10 kGy was introduced. Good results have been obtained in the Netherlands. Moreover, it has been shown that with higher doses the development of spores is clearly restrained. Changes in etheric oil content or taste, which for spices is very important, have so far not been observed with irradiation (69). As a replacement for gassing, and on the grounds of experience gained with tests done in the Netherlands and in Hungary, irradiation of spices should also be given priority. In this respect, the irradiation of grains, nuts, cocoa-beans, etc., should also be considered, in order to eliminate mould growth and possible production of carcinogenic mycotoxins (Table 2).

Table 2. Recent approvals (1967-1977) for use of low-dose irradiation (source: Vas, ref. 53).

Food item	Approved dose (kGy)	Location
Fresh red meat	6 - 8	USSR
Processed red meat	6 - 8	USSR
Chicken[1]	3	Netherlands
Chicken	7	Canada
Chicken	6	USSR
Chicken[1]	2 - 7	WHO
Spices	8 - 10	Netherlands
Spices	5	Hungary

1. Unlimited clearances for human consumption, other clearances are for experimental batches, for marketing test and consumer acceptance.

Finally, the irradiation of animal feeds should be briefly mentioned. Feeds from both animal and vegetable origins have played a considerable role for many years in the contamination cycle of <u>Salmonella</u>. Pelleting of meals can reduce the number of non-spore-forming bacteria by 2 to 3 logs. Irradiation of feeds has been shown to be effective for microbiological as well as zootechnical aspects (70, 71, 72).

A combined process of pelleting and irradiation could have special advantages in regard to post-contamination which, with irradiation of packed feeds, can be prevented. Irradiation alone could save energy costs which are considerable in steam-pressure pelleting.

In retrospect, one wonders why food irradiation, considering all the many favourable results, is either not applied, or only hesitantly so, in practice. It is mainly the emotional resistance of the consumer against all forms of nuclear energy that is to blame. Moreover, and also emotional, there are many misconceptions regarding food irradiation, beginning with the use of incorrect terminology by the media such as "radioactive irradiation" and "radioactive food" and ending with confusing the problems surrounding food irradiation with the atomic source. One may wonder whether specialists in the field, as the most qualified members of society to explain and correct these misunderstandings, having fulfilled their task in the past.

It is to be hoped that the information given here will help to correct this omission, so that in the future, wherever it is possible and desirable, a method of food preservation can be used that is effective and without threat to human health.

5. REFERENCES

1. Abdussalam, M., Fleischwirtschaft 60 (1980) 1212.
2. Grossklaus, D., Fleischwirtschaft 60 (1980) 1141.
3. Heden, C.G., NIPH Annals 3 (1980) 15.
4. Aserkoff, B., Schröder, S.A. & Brackman, Ph.S., Am.J. Epidemiol. 92 (1970) 13.
5. Levy, B.S. & McIntire, W., J.Am.med.Assoc. 230 (1974) 1281.
6. Food Irradiation in the USA, Report prepared for the inter-departemental Committee on Radiation Preservation of Food, December 1978.

7. Kampelmacher, E.H., In: Proc.SOS/70 3rd International Congress of Food Science and Technology (Washington, D.C. 1970) 681.
8. Kampelmacher E.H., "Bacterial indicators. Health hazards association with water", Special Technical Publication of the American Society for Testing and Materials (1977) 635 148.
9. Bryan, F.L., J.Food Prot. 43 (1980) 140.
10. Edel, W., Schothorst, M. van & Kampelmacher, E.H., In: Proceedings of the International Symposium on Salmonella and Prospects for Control (Guelph, Canada, 1978).
11. Bischoeff, J. & Rohde, R., Berl.Münch.tierärztl.Wschr. 69 (1956) 50.
12. Müller, J., Bull.Off.Int. Epizoot. 48 (1957) 323.
13. Jacobs, J., Guinee, P.A.M., Kampelmacher, E.H. & Keulen, A. van, Zbl.VetMed. B 10 (1963) 542.
14. Velaudapiilai, T. & Nitiananda, K., Kamini Meedeniya, Zschr. Hyg. 149 (1963) 122.
15. Craven, P.C., Mackel, D.C., Baine, W.B., Barker, W.H., Gangarosa, E.J., Goldfield, M., Rosenfeld, H., Hetman, R., Lachapelle, G., Davies, J.W. & Swanson, R.C., Lancet i (1975) 788.
16. Gottschalk, H.M., Food Irrad.Inf. No 7 (1977).
17. Tamminga, S.K., Beumer, R.R. & Kampelmacher, E.H., J.Hyg., Camb. 76 (1976) 41.
18. Tamminga, S.K., Beumer, R.R. & Kampelmacher, E.H., J.Hyg., Camb. 80 (1978) 143.
19. Schothorst, M. van & Kampelmacher, E.H., J.Hyg., Camb. 65 (1967) 321.
20. Schothorst, M. van, Northolt, N.J., Kampelmacher, E.H. & Notermans, S., J.Hyg., Camb. 76 (1976) 57.
21. Fantasia, L.D., Presented at the 74th Annual Meeting of the American Society for Microbiology (Chicago, Ill., 1974).
22. Brown, L.D. & Dorn, R.R., J.Food Prot. 40 (1977) 712.
23. Todd, E.C.D., J.Food Prot. 43 (1980) 129.
24. Siliker, J.H., J.Food Prot. 43 (1980) 307.
24a Kuipschild, P., Tijdschr. soc. Geneesk. 59 (1981) 356.
25. Mossel, D.A.A. & Hoogendoorn, J., Industr.Med. 40 (1971) 25.
26. Report of a WHO Working Group, "Aviation Catering" (Torremolinos 1976). Rep.Office WHO Copenhagen (1977).
27. Todd, E.C.D., J.Food Prot. 41 (1978) 559.
28. Vernon, E., Publ.Hlth., Lond. 91 (1977) 225.
29. Bryan, F.L. & McInley, T.W., J.Milk Food Technol. 37 (1974) 420.
30. Pace, P.J. & Krumbiegel, E.R., J.Milk Food Technol. 36 (1973) 42.
31. Lynt, R.K., Kautter, D.A. & Read, R.B., J.Milk Food Technol. 38 (1975) 546.
32. Huss, H.H., Appl.environm.Microbiol. 39 (1980) 764.
33. Notermans, S., Dufrenne, J. & Oosterom, J., Appl.environm. Microbiol. (1981, in press).
34. Johnson, H.C., Baross, J.A. & Liston, J., J.Am.vet.med.Ass. 159 (1971) 1470.
35. Barrow, G.I., Postgrad.med.J. 50 (1974) 612.

36. Sakasaki, R.T., Karashimada, T., Yuda, K., Sakai, S.,
    Asawaka, Y., Yamazaki, M., Nakanishi, H., Kobayashi, K.,
    Nishio, T, Okazaki, H., Doke, T., Shimaka, T. & Tamura,
    K., Arch.LebensmittHyg. 30(1979) 103.
37. Butzler, J.P. & Skirrow, M.B., Acta paediatr.belg. 32
    (1979) 89.
38. Report of a WHO Expert Committee, "Parasitic Zoonoses",
    Techn. Rep. Series 637 (1979).
39. Smith, J.W. & Wootten, R., In: Advances in Parasitology,
    Vol.16, p.93. Academic Press, London, 1978.
40. Goresline, H.E., Ingram, M., Machuch, P., Moquot, G.,
    Mossel, D.A.A., Niven Jr., C.F. & Thatcher, F.S., Nature
    204 (1964) 237.
41. Ingram, M. & Farkas, J., Acta aliment. 6 (1977) 123.
42. Hobbs, B.C., Reeves, J.C., Gurside, J.S., Gordan, R.F.,
    Barnes, E.M., Shrimpton, M.A. & Anderson, E.S., Hlth Bull.
    Min. Hlth Lab. Serv. 19 (1960) 178.
43. Diehl, J.F., Zbl.Bakt.Hyg., I.Abt. Orig. B 156 (1972) 157.
44. Report of a Joint FAO/IAEA/WHO Expert Committee, "Whole-
    someness of Irradiated Food", WHO Techn. Rep. Series 604
    (1977)
45. Mossel, D.A.A., Schothorst, M. van & Kampelmacher, E.H.
    In: Elimination of harmful organisms from food and feed
    by irradiation. IAEA Vienna (1968) 43.
46. Mossel, D.A.A., J. Food Qual. 1 (1977) 85.
47. Rowley, D.B., Anellis, A., Wierbicki, E. & Baker, A.W.,
    J.Milk Food Technol. 37 (1974) 86.
48. Thomson, J.E., Bailey J.S. & Cox, N.A., Poultry Sci. 58
    (1979) 139.
49. Snijders, J.M.A., Schoenmakers, M.J.G., Gerats, G.E. &
    Pijper, F.W., Fleischwirtschaft 59 (1979) 656.
50. Mulder, R.W.A.W., Notermans, S. & Kampelmacher, E.H.,
    J.appl.Bact. 42 (1977) 179.
51. Urbain, W.M., Food Irrad. Inf. No. 8 (1978) 14.
52. Notermans, S., Schothorst, M. van, Leusden, F.M. van &
    Kampelmacher, E.H., Tijdschr.Diergeneesk. 100 (1975) 648.
53. Vas, K., General survey of irradiated food products
    cleared for human consumption in different countries.
    Joint FAO/IAEA/WHO Advisory Group on International Acceptance
    of Irradiated Food, GA-143/INF/2. IAEA, Vienna, 1977.
54. Ley, F.J., Kennedy, T.S., Kawashima, K., Roberts, Diane &
    Hobbs, Betty, J.Hyg., Camb. 68 (1970) 293.
55. Simard, C., Lachance, R.A. & Moreau, R.R., Can.Inst.Food
    Sci. Technol. J. 6 (1973) 250.
56. Billon, J. & de la Sierra Serano, D., Food Irradiation 8
    (1968) 22.
57. Beumer, R.R., Tamminga, S.K. & Kampelmacher, E.H., Micro-
    biological investigation of "Filet Americain", Arch. Lebens-
    mittHyg. 1982 (in press).
58. Gomberg, H.J. & Gould, S.E., Science 118 (1953) 75.
59. Smith, P.H., USDA Meat and Poultry Inspection Program.
    Agric. Inf. Bull. No 377 (1975).
60. Dossow, J.A. & Miyarichi, D.T., In: Radiation preservation
    of foods. Natn.Res.Council Publ. No 1273 (1965) 53.
61. Pelroy, Gretchen A. & Seman, Jr.J.P., J.Milk Food Technol.
    31 (1968) 231.

62. Ehlermann, D. & Munzner, R., Arch.LebensmittHyg. 27 (1976) 41.
63. Gonzales, O.N., Report No R-1737-F, IAEA, Vienna, 1979.
64. Anonymus, Brit.Med.J. No 6026 (1976) 1.
65. Wahba, A.H.W., Bundesgesundh.Bl. 19 (1976) 139.
66. Ayres, P., Hlth Hyg. 3 (1979) 11.
67. Hobbs, G., Food Irrad.Inf. No (1977) 39.
68. Anonymus, Wkly Epid.Res. 42 (1974) 351.
69. Zehnder, H.J., Alimenta 18 (1979) 43.
70. Edel, W., Guinee, P.A.M., Schothorst, M. van & Kampelmacher, E.H., Zbl.VetMed. B 14 (1967) 393.
71. Reusse, U., Bischoff, J., Fleischauer, G. & Geister, R., Zbl.VetMed. B 23 (1976) 158.
72. Dammers, J., Kampelmacher, E.H., Edel, W. & Schothorst, M. van, Food Irradiation. IAEA, Vienna (1966) 159.

Irradiation of foodstuffs — Technological aspects and possibilities

D.Is. Langerak

1. INTRODUCTION

It is a well-known fact that all foodstuffs are subject
to physiological, chemical, biochemical and microbial decay.
In general, the most common form of deterioration of foodstuffs
is that caused by microorganisms; but enormous losses are also
caused by desiccation, so adequate packaging is very important.

Various preservation methods have been developed over the
years in order to prevent microbial decay.

Recently, in addition to the usual methods such as heating,
cooling, addition of chemicals, etc., ionizing radiation has
also been used, especially in the medical industry. This method
of preservation is gaining more and more ground in the food
industry because the application of radiation offers certain
prospects. Every preservation method has its own specific
properties and applications. This applies also to preservation
by means of ionizing radiation. This can best be illustrated
by discussing some packaging methods.

Most preservation methods require specially adapted packaging;
for sterilization by heating, metal packaging (tins) with
limited dimensions are usually used for rapid transfer of heat;
for cooling, open packaging is required to allow for quick
penetration of cold into the product. In the case of radiation
a closed packaging can be applied with dimensions which are
considerably larger than those required for a heat treatment.

Food Irradiation Now — ISBN 90-247-2703-0
Copyright Gammaster and Martinus Nijhoff/Dr W. Junk Publishers

2. PROPERTIES OF RADIATION

For food preservation and irradiation of non-foods only gamma rays and electrons are used, produced by Cobalt-60 sources and electron generators respectively. Gamma rays form part of the electro-magnetic spectrum, which is shown in Fig. 1. In this spectrum the waves of various kind of rays become shorter from left to right.

In addition to UV rays there are röntgen (X) and gamma rays. The shorter waves, the more penetrating the radiation, due to the higher energy (often expressed as eV). Whereas UV rays cannot penetrate glass (they are absorbed) X-rays and gamma rays penetrate both glass and packaging. The penetrating qualities of gamma rays through packaging are dependent on the following factors:
- the energy of rays;
- the specific mass of the packing material;
- the density of the packed product.
The density of the packed product is usually the most important factor. If the density is 1, then half the gamma rays will be absorbed with a product thickness of 11 cm. With a density of 0.5, this value is approximately 22 cm. The dimensions of the packaging intended for an irradiation treatment also depend, in addition to the density of the packed product, on the required uniformity of radiation dose in the packaging: the D_{max}/D_{min} ratio, as shown in Fig. 2. D_{max} is the highest absorbed dose in the packaging and D_{min} the lowest (1).

Uniformity is increased by irradiating the product on two sides. If the product requires great uniformity of radiation dose, i.e. small D_{max}/D_{min} ratio, then the dimensions of the packaging must be adapted. If the product allows a large D_{max}/D_{min} ratio, e.g. 3, as is the case when onions are irradiated to prevent sprouting, then it is possible to use bulk bins with a content of 1 m^3 measuring 1 m x 1 m x 1 m. The required uniformity is determined by the sensitivity of the product to irradiation. In the case of a large difference in D_{max} and D_{min} non-acceptable taste, odour and colour differences

42

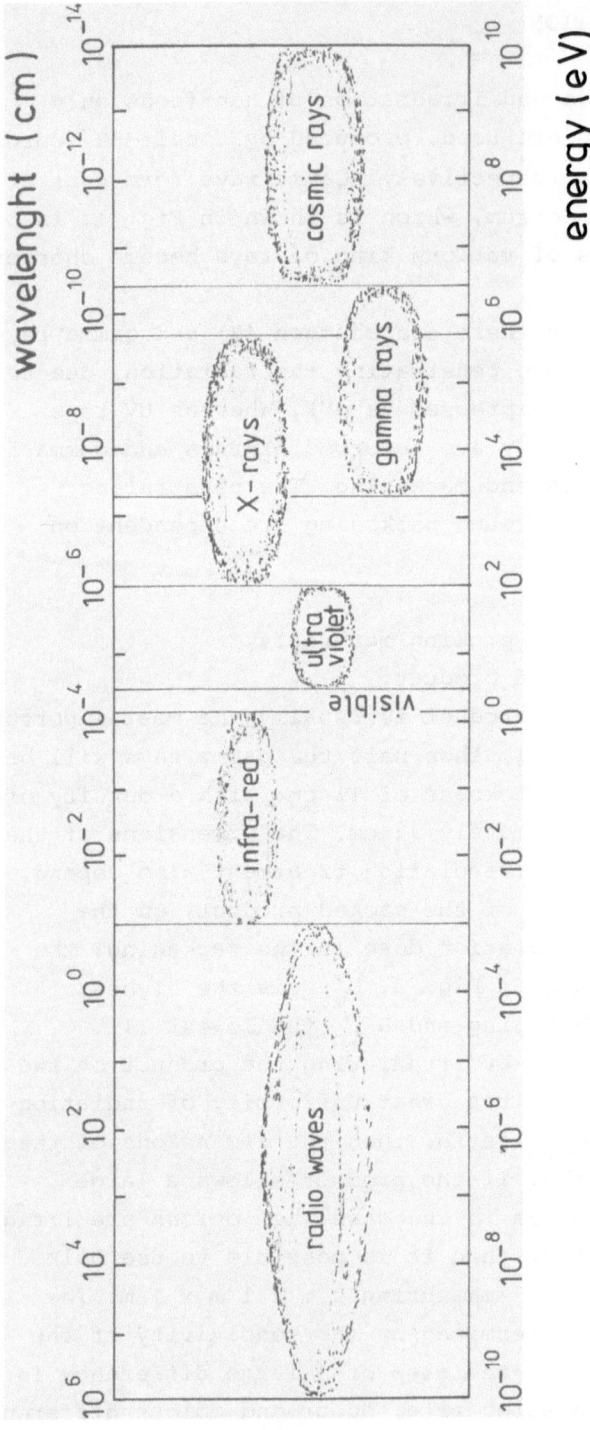

FIG. 1. Electro-magnetic spectrum.

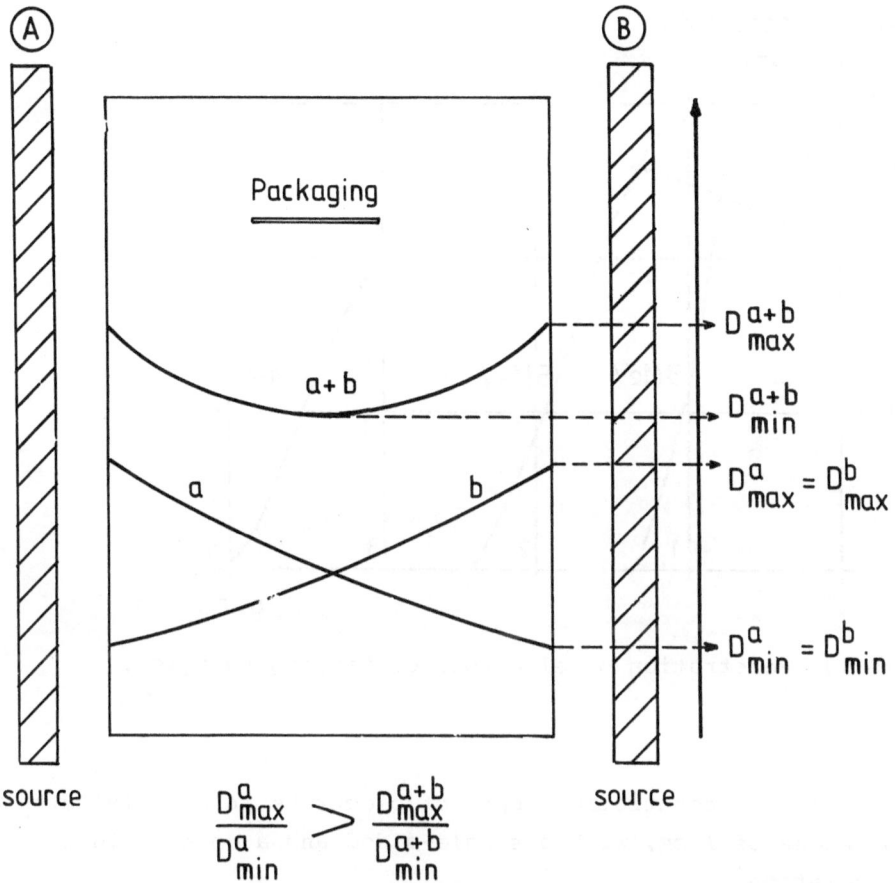

FIG. 2. Penetration curve and dose curve of a double-sided gamma irradiation.

can arise in this same packaging. Because of this, irradiation in pallet packaging is not always possible. The D_{max}/D_{min} relationship is, therefore, related to the quality requirements of the product.

Not only gamma rays but also electrons are used in food preservation. In general they are less penetrating than gamma rays and are, therefore, only used for products of limited thickness or for surface irradiation. The penetration of electrons into the material takes place according to the Bragg curve (see Fig. 3).

relative dose (%)

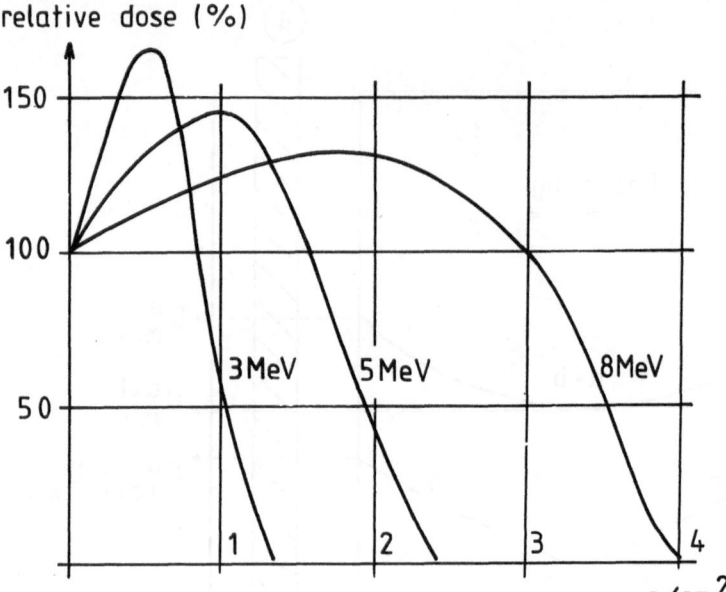

FIG. 3. Penetration of electrons of varying energies.

Table 1. The penetration depth of electrons in material with a thickness of 1 cm, with a single-sided and a double-sided irradiation

Energy (MeV)	Maximum depth (cm)	Effective packaging thickness (cm)	
		single-sided	double-sided
1	0.5	0.3	0.9
2	1.0	0.6	1.7
4	2.0	1.2	3.5
6	3.0	1.9	5.1
8	4.0	2.5	7.0
10	5.0	3.1	8.9

The penetration of electrons is also determined by the
energy of the rays and the density of the material. Because
dose distribution runs from 100 to 0%, the concept of effective
packaging thickness is incorporated; this is 3/5 of the maximum
depth of penetration. With a double-sided irradiation, this
packaging thickness is increased (2). An outline of the penetration depth of electrons with varying energies is given in
Table 1.

Table 1 shows that, in comparison with gamma rays, the
packaging thickness with electrons, even with high energies,
is limited. For bulk packaging, therefore, only gamma rays
are suitable.

2.1 Dose

As with heating or cooling treatments, a unit is used for
the dose of radiation given. The former unit was the rad
(radiation absorbed dose): 1 rad = 100 erg absorbed energy
per gram of tissue; 1000 rad = 1 krad; 1000 krad = 1 Mrad.
Since the adoption of the new Sl units, the Gy (gray) is used:
1 Gy = 1 Joule/kg = 100 rad.

2.2 The effect of radiation

As far as food preservation is concerned, ionizing radiation
has the following three effects
1. (Micro)biological effect. Radiation can destroy (micro)
organisms. Depending on the dose, radiation can
- desinfest (inactivation of insects, larvae, eggs);
- decontaminate (pathogen-free: elimination of non-sporulating
pathogenic microorganisms);
- pasteurize (lowering of the bacterial count, shelf-life
extension);
- sterilize (destruction of all microorganisms, more or less
unlimited shelf-life).
2. Physiological effect. By influencing biochemical processes
in the product, radiation can produce the following effects:
- inhibition of sprouting (potatoes, onions, carrots);
- delay of growth and ripeness (mushrooms, fruits).

3. Physical effect. Changes in permeability (shortening of
drying and cooking times in dried vegetables).

How radiation brings about the above-mentioned effects is
very complicated. It is assumed that the absorbed energy produces
very small molecular changes in tissue, whereby certain biochemical processes can be decreased or increased.

How does irradiation influence the shelf-life of fruits
and vegetables compared with other preservation methods?
This can be shown with the help of Table 2 which compares the
effects of cooling, heating, irradiation and food additives
on components which govern the shelf-life and quality of fruits
and vegetables. These components are:
- the number of microorganisms in and upon the product;
- biochemical processes in the product (senescense, respiration);
- desiccation.

Table 2 shows that cooling does not kill microorganisms,
but only delays their growth. The rate of senescence (dissimilation) is delayed by cooling according to the Q_{10} rule.
Cooling, however, promotes desiccation because open packaging

Table 2. Effect of various preservation methods on the components
governing shelf-life and quality.

Components	Preservation methods			
	Cooling[1]	Heating	Radiation[2]	Additives
Microorganisms	±	+	+	+
Senescence	+	+	±	0
Desiccation	−	−	+	0
Product	0	−	0	0 (±)

+ = positive; - = negative; ± = moderate; 0 = no effect.
1. Cooling to 0 °C.
2. Dose lower than 5 kGy.

is necessary to allow the cold to penetrate. The fresh character of a product is maintained by cooling. A heat treatment kills the microorganisms, arrests senescence by the inactivation of enzymes, and promotes desiccation (loss of weight). The product is completely changed by a heat treatment; the fresh character disappears.

Like heating, an irradiation treatment inactivates most of the microorganisms. It also has a delaying effect on senescence (ripening) by reducing the enzyme activity in the product. The effect of irradiation on senescence is, however, less radical than that of cold or heat. A combination of irradiation with a moderate cooling treatment is, therefore, usually necessary.

No extra desiccation appears with irradiation; it is even possible to irradiate a fresh product in sealed packaging. In such packaging, a modified air composition (low O_2 and high CO_2 content) can develop from the respiratory activity of the product. This kind of gas storage delays senescence, thus giving a longer shelf-life. Like cooling, an irradiation treatment does not change the fresh character of the product (no temperature increase occurs!). So far, no harmful substances have been shown in products which have been preserved by irradiation.

Food additives diminish the number of microorganisms, but generally have no effect on senescence, on desiccation or on the product itself. The application of chemicals, however, leaves residues in and on the product, which can be harmful.

3. POSSIBILITIES OF IRRADIATION

It has been shown that irradiation offers a number of specific possibilities in food preservation, because of the preserving effect without considerable temperature increase. An absorption dose of 10 kGy per hour is equal to 10^4 J kg^{-1} h^{-1}, resulting in an increase in temperature of 2.78 $^\circ$C.

In the past, irradiation has often been applied as a substitute for cooling. Irradiation, however, requires its own specific technology. In fact, irradiation should be considered as being

48

Table 3. A summary of radiation applications in food preservation.

Product	Dose (kGy)	Aim
Tuber, bulb and root vegetables	0.02-0.15	inhibition
Grains, grain products dried fruit	0.2-0.5	desinfestation
Pome and stone fruit, tropical fruit	0.25-1.0[1]	delay of rotting, ripening and storage disease
Pre-packed vegetables	0.5-2.0	prolonged shelf-life
Soft fruit	2.0-2.5	prolonged shelf-life
Canned products	2-10	sterilization (radiation + heat)
Deep-frozen, dried products (raw material)	5-10	decontamination
Non-food	10-50	sterilization

1. Combined with heat (40-55 $^{\circ}$C).

supplementary to or an improvement of an existing preservation method. It may only totally or partially replace such a method (e.g. refrigeration) if the product does not tolerate that particular treatment (alternative).

Over the years a wide range of possibilities has been developed by irradiation research (see Table 3).

4. CONDITIONS FOR A CORRECT IRRADIATION TREATMENT

To achieve optimal results with the application of irradiation it is desirable to know the marginal conditions. For "living" products such as agricultural and horticultural products, the following factors are important.
- The initial quality of the product (irradiation cannot

convert a bad product into a good one).
- Ripening stage: the ripening of a product can only be delayed
by irradiation when the fruit is in the climacterium. In controlling decay, irradiation of ripe fruits is more successful,
because in unripe fruits the natural resistance to decay can
be reduced by irradiation.
- The initial contamination: to apply the optimal dose
knowledge of type and level of the contamination is important.
From Table 4 it can be seen that microorganisms have different
resistances with regard to irradiation. This resistance is
expressed as a D_{10} value (the irradiation dose corresponding
with a decimal reduction). When level and type of contamination
are known, it is theoretically possible to calculate a dose
at which all microorganisms are inactivated.
- Time of irradiaton: for physiological and microbiological
reasons the product should be irradiated as soon as possible
after harvesting. Investigations have shown that, with ageing
of the product, irradiation is less effective. Also, after

Table 4. Decimal reduction (D_{10} value) for some microorganisms.

Species	Dose (kGy)
Pseudomonas spp.	0.10 - 0.20
Escherichia coli (aerobic)	0.12 - 0.35
Escherichia coli (anaerobic)	0.20 - 0.45
Salmonella spp.	0.20 - 0.50
Streptococcus faecalis	0.50 - 1.00
Fungus spores (Penicillium, Aspergillus, etc.)	0.50 - 0.70
Bacillus pumilus (spores)	ca. 1.70
Clostridium sporogenes	1.60 - 2.20
Clostridium botulinum	1.50 - 2.50
Micrococcus sodonensis	ca. 1.95
Micrococcus radiodurans	>5.00

harvesting, the number of microorganisms increases very rapidly
and it is also possible that the microorganisms become more
resistant with storage resulting in the necessity of a higher
dose, with all the consequences.
- The packaging must be adapted to the irradiation procedure.
Because irradiation does not prevent desiccation, sealed
packaging is usually recommended; the product is then also
protected against external influences (re-infection).
- Storage and transport conditions. These must also be adapted
to both the product and the irradiation procedure. In combination
with irradiation a too low storage temperature can stimulate
tissue damage.

In "dead" products, such as raw materials, in which few
or no enzyme activities are present, the moisture content in
the product plays an important role. In the case of a low
moisture content, the effect of irradiation on the product
and the inactivation of microorganisms is less than when the
moisture content is high. In general, a high O_2 content promotes
the inactivation of microorganisms by irradiation. However,
under anaerobic conditions a higher dose is required for the
same microorganisms.

5. TECHNOLOGICAL ASPECTS

It is impossible to discuss these aspects for all products;
a few examples will be given for some fields of application.

5.1 Desinfestation
Grain and grain products, especially from tropical countries,
are often infested with insects, larvae and eggs. This causes
enormous losses and the countries importing the grain demand
stringent quality requirements. Controlling these pests by
means of insecticides is possible. They are usually introduced
into the silos or sprayed onto products packed in jute sacks.
This, however, means that re-infestation can occur. The ideal
packing for grain and grain products is polypropylene, because
it cannot be eaten by insects and the product itself is also

protected from damp, etc.

Because these bags are impermeable to gases they cannot be gassed with insecticides, but can be treated with gamma rays because these rays can penetrate straight through the packing and the product. Because the packing remains sealed, no re-infestation can occur.

Another advantage of irradiation is that it can control moulds which sometimes cause rotting and produce mycotoxins that are dangerous to health. Desinfestation by radiation is also a "clean" treatment; it leaves no residues as is the case with insecticides.

5.2 Decontamination

Frozen products from animal origin are often contaminated with pathogens such as Salmonella.

Frozen vegetables and fruit are often contamined with yeasts and other bacteria that cause rotting. Because of the more stringent quality requirements, these contaminated products are no longer accepted for either export or consumption. Decontamination is, therefore, essential. Until now the only possibility was heating, but the product loses its fresh character and the quality of the product is seriously diminished (e.g. consistency, colour and taste). Also, heating requires much energy, because the product has to be thawed and re-frozen.

In addition to deep-frozen products, microbial contamination also occurs in dried products such as fruit powders, dried vegetables, herbs, spices, and raw materials.

These products form a focus of infection during further processing. Decontamination by ethylene oxide is possible, but this is toxic, leaves residues in the product and is forbidden in many countries. Decontamination by gamma rays is, therefore, the simplest solution. An example of this is given in Table 5.

When irradiating frozen products it is important that the temperature does not rise too much during the treatment. Temperature measurements have shown that the irradiation time should not exceed 1 to 1½ hours, because otherwise the temperature will rise above -15 $^{\circ}$C. Packaging in polystyrene containers

Table 5. Effect of irradiation on the microbial quality of non-blanched silverskin onions. Average count per gram of 3 samples.

Dose (kGy)	Viable count	Entero-bacteriaceae	Escherichia coli	Staphylococcus aureus	Yeast and moulds
0	7.3×10^6	3.2×10^4	neg.	<10	2.4×10^3
0.9	1.7×10^6	2.0×10^2	neg.	<10	2.4×10^3
1.4	3.8×10^6	3.7×10^1	neg.	<10	<10
1.8	4.4×10^6	<10	neg.	<10	<10
2.7	2.8×10^6	<10	neg.	<10	<10

or using a polystyrene lining (1 cm thick) in the containers can considerably delay the temperature rise.

In relation to decontamination, irradiation offers the following advantages:
- treatment in the original packaging
- no re-contamination
- maintenance of fresh or deep-frozen character because there is no temperature rise during irradiation (cold pasteurization)
- no loss of quality
- no residues, as with gassing
- saving of material (no re-packing)
- saving of energy.

5.3 Pasteurization

Most horticultural products deteriorate during storage, transportation and distribution/marketing, due to microbial decay and desiccation. Irradiation can control microbial decay, but not desiccation.

Losses due to desiccation can reach 10 to 30%. This weight loss can largely be prevented by storage in high humidity (Fig. 4). Because during transportation and marketing the humidity is often low, adequate packaging of horticultural

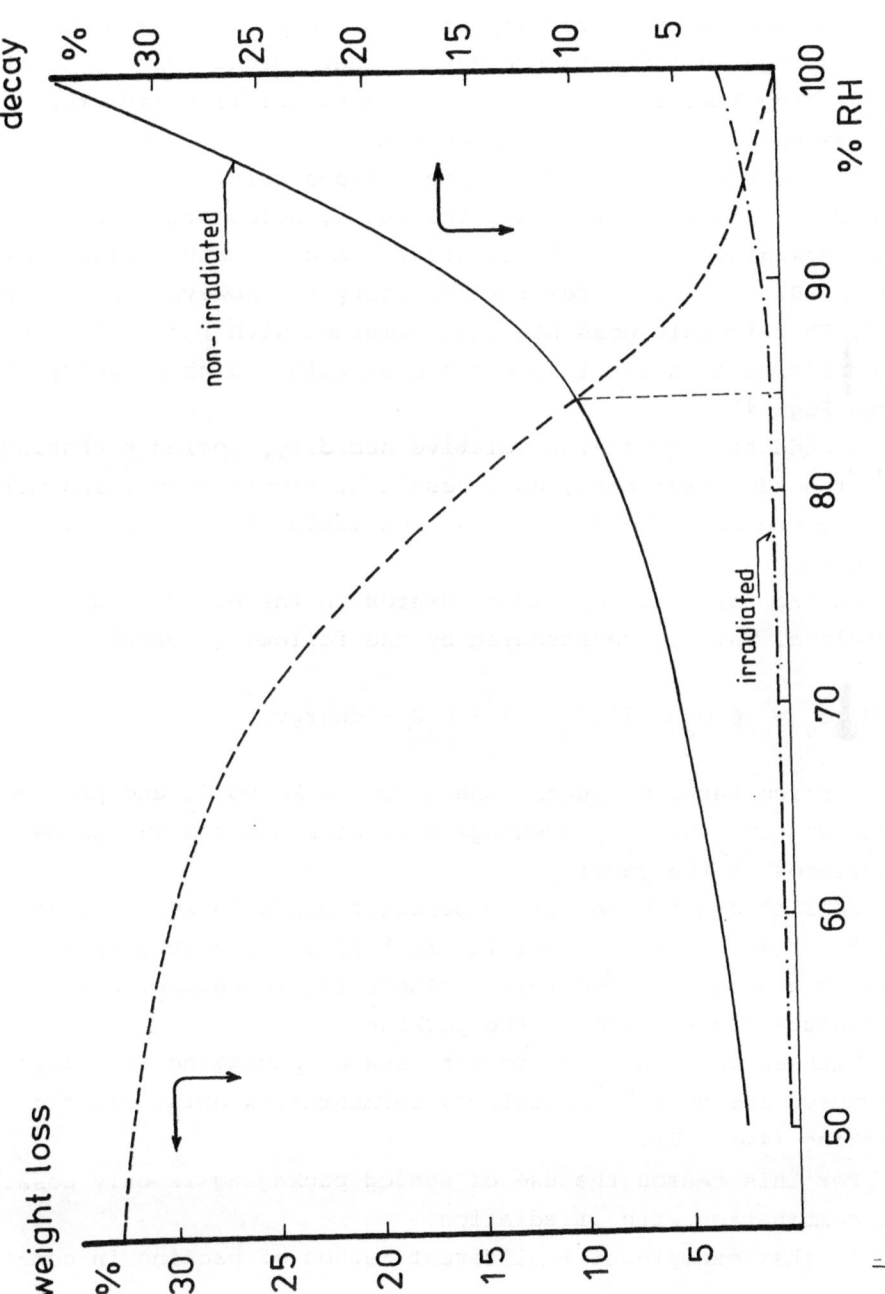

Fig. 4. Effect of relative humidity and irradiation on weight loss and decay of vegetables.

products is very important. Various plastics can be used, such
as polythene, polypropylene, polyvinyl chloride (pvc), polystyrene and polyvinylidene chlorides (pvdc). The most commonly
used foil is polythene (low density). Due to the low permeability
of the packing and water vapour from the product, a microclimate
with a high humidity is present so that weight loss is limited.
But a high humidity also has disadvantages: it stimulates,
for example, decay by microorganisms.

Fig. 4 shows the relationship between relative humidity
and decay. To prevent decay, therefore, holes are made in
the packaging so that the relative humidity cannot rise above
85 to 90% The ideal relative humidity is, however, more than
95%. This is only possible when combined with radiation, because
then microbial decay is reduced even with a high humidity
(see Fig. 4)

In addition to a high relative humidity, sealed packaging
has the advantage that, as a result of respiration (dissimilation)
of the product, the air composition inside the packing is
altered.

During respiration, carbohydrates in the planttissue are
oxidized. This is represented by the following formula:

$$C_6H_{12}O_6 + 6 \ O_2 \leftrightarrows 6 \ CO_2 + 6 \ H_2O + energy.$$

Horticultural products, therefore, take up O_2 and produce
CO_2, whereby the O_2 percentage decreases and the CO_2 percentage
increases in the packing.

This alteration in air composition has a delaying effect
on the respiration process (ageing) of the product, thus
giving a prolonged shelf-life. There is, therefore, a sort of
miniature gas storage in the packing.

With an unirradiated product, sealed packaging is impossible
because, due to microorganisms, fermentation exists in the
packing (4, 5, 6).

For this reason the use of sealed packaging is only possible
in combination with irradiation.

Another example of a different method of packing in combination

with irradiation, is the packing of soft fruit such as strawberries, raspberries and blackberries.

Soft fruit generally has a limited shelf-life because it
is very susceptible to desiccation, transportation damage and
rotting. Closed packaging offers good protection but is impossible
because the high humidity causes rapid rotting. Only in
combination with radiation can closed packaging be used, with
the following effects:
- limited desiccation (green calix, fresh appearance)
- better quality (less damage and rotting)
- prolonged shelf-life

An improved method of packing combined with radiation can
also be applied to other products such as meat and cooked
meats. These products also deteriorate in quality due to
weight loss, decay and discolouration. A suitable packing is
a flexible vacuum packaging with a low permeability for O_2 and
water vapour. Several laminates are available for this purpose.
This vacuum packaging, however, gives problems due to infection
by both lactobacilli and anaerobic bacteria, which multiply
rapidly in the anaerobic environment, causing decay.

Combined with an irradiation treatment, however, vacuum
packaging can be used because the bacteria are eliminated by
radiation. In this way, a prolonged shelf-life is achieved;
the vacuum and water-vapour-proof packing limits discolouration
and desiccation.

5.4 Sterilization

Metal packaging is usually used for the sterilization of
food. The dimensions depend on whether the product belongs to
the convection or conduction type. With the conduction type,
especially, the dimensions must be limited, otherwise the
product has to be heated for too long in order to achieve
the desired sterility.

Flexible, synthethic materials have also been used for
sterilization for some years now, the so-called retort
packaging, mainly consisting of laminates of nylon, aluminium and
polypropylene. Because heat sterilization occurs at 121 $^{\circ}$C,

the qualities required of the plastic packaging are very high;
the number of plastics suitable for heat sterilization is,
therefore, limited.

Combined with radiation, however, it is possible to reduce
the heat treatment, expressed as an Fo value (Fo value is the
process value = number of minutes necessary to sterilize a
product at 121 $^{\circ}$C), by 50% of the original value. This results
in either a shorter heat treatment, or sterilization at a
lower temperature, e.g. 100 $^{\circ}$C (8). Table 6 gives a clear
view of a shorter heat treatment.

Because of these shorter or lower heat treatments less
stringent demands are made upon the plastic material. The
combination of heat and irradiation for the sterilization of
foodstuffs offers the following perspectives:
- packaging in large units, especially important for conduction
types;
- greater choice of packing material, which means it can be
cheaper;
- saving of fossil energy because the treatment is either

Table 6. Effect of a combined treatment on the Fo value (see
text) and quality with regard to a single heat treatment.

Product	Heat Fo value	Heat + irradiation		Quality improvement[1]		
		Fo value	dose (kGy)	colour	taste	texture
Strawberries	3.2 min/ 85 $^{\circ}$C	0.9 min/ 85 $^{\circ}$C	< 1	+	$\pm$	+
Pears	10.0 min/ 85 $^{\circ}$C	7.2 min/ 85 $^{\circ}$C	< 3	+	$\pm$	+
Asparagus	19.3 min/115 $^{\circ}$C	8.0 min/115 $^{\circ}$C	< 3	+	$\pm$	+
Spinach	4.7 min/121 $^{\circ}$C	0.04 min/121 $^{\circ}$C	> 3	$\pm$	+	$\pm$
French beans	17.5 min/121 $^{\circ}$C	7.4 min/121 $^{\circ}$C	< 3	$\pm$	+	+
Peas	18.0 min/121 $^{\circ}$C	6.0 min/121 $^{\circ}$C	$\leq$ 5	$\pm$	$\pm$	+

1. + = clear; $\pm$ = slight.

shorter or is done at a lower temperature;
- better quality of the product.

5.5 .Disadvantages of irradiation

As every preservation method, food irradiation has disadvantages as well as advantages. When too high a dose is given, texture and colour losses can occur. Also, some aromas and flavourings are sensitive to radiation so that an "off-flavour" can occur that is analogous to a "deep-freeze taste" or a "tin taste". This can occur especially in foods with a high protein or fat content.

Where these difficulties could exist, radiation is done in combination with other preservation methods, for example with moderate cooling, a mild heat treatment or under vacuum. Colour, taste and odour changes can also be prevented by irradiation under vacuum or at a very low temperature (e.g. -80 $^{\circ}$C). These two methods are applicable for fish, meat, meat products and other radio-sensitive products.

A combination of a mild heat treatment with irradiation is very effective in controlling moulds on products which are very sensitive to heat or high doses.

The above-mentioned combination gives a synergistic effect: the combination has a greater effect on the inactivation of fungi than the sum of the single treatments (see Fig. 5). Consequently, a lower heat treatment and a lower irradiation dose can be applied with no adverse effects on the product.

The development and application of this combined treatment is still being intensively studied. However, practical applications have already been started with tropical products such as mangoes and papayas (10).

Instead of a fungicide, a combined heat and irradiation treatment is used, which prevents decay and delays ripening. In the case of the mango, the mango weevil is also controlled at the same time. Furthermore, this combined treatment now offers the possibility of transporting the product by ship instead of by aeroplane, resulting in a lowering of transport costs by a factor 3.

58

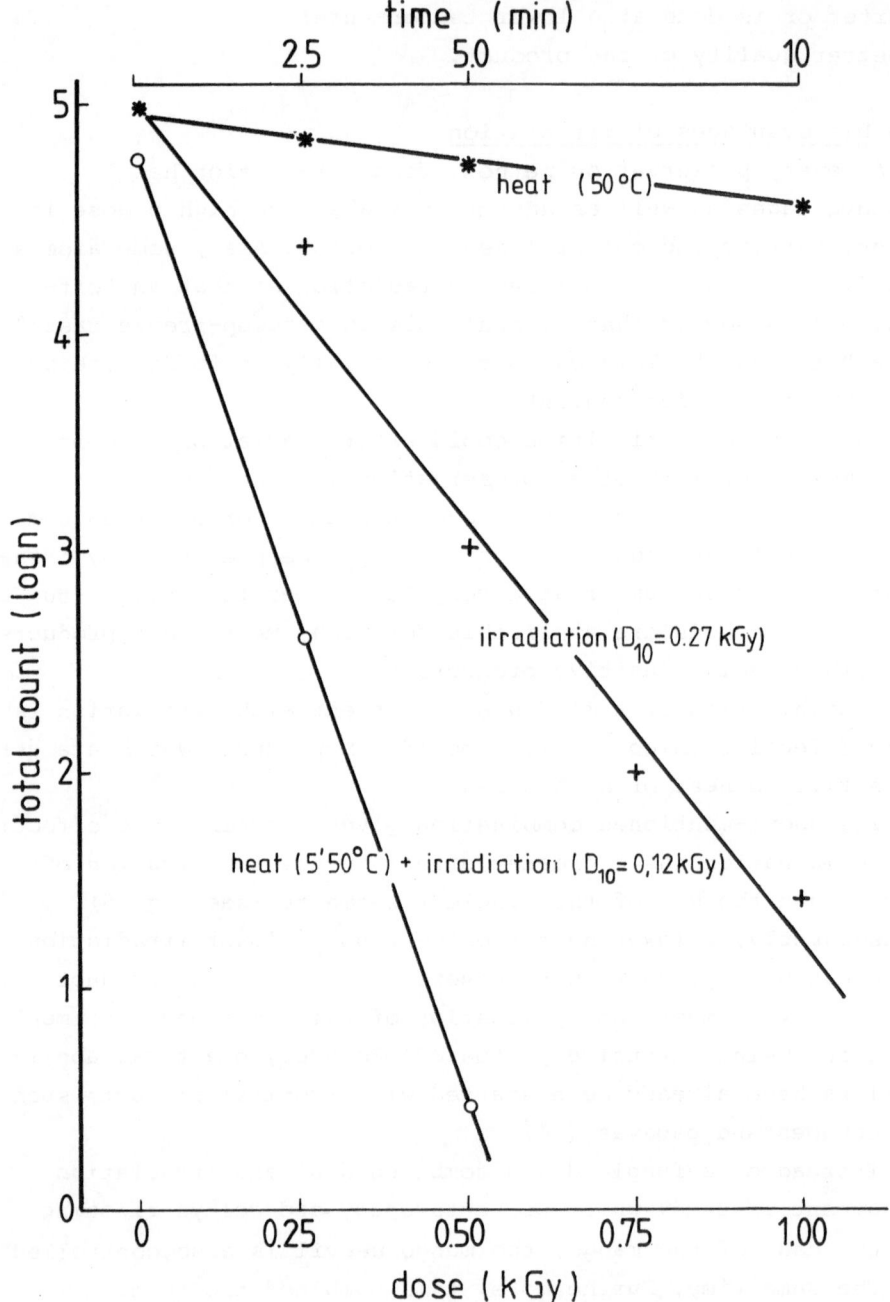

FIG. 5. Effect of heat (50 $^{\circ}$C), irradiation and a combination of both on a D_{10} value of <u>Penicillium expansum</u> spores.

6. CONCLUSIONS

This review shows that irradiation has its restrictions, but also a wide spectrum of application possibilities. We can, therefore, expect irradiation to claim its place in the field of food preservation, especially where the conventional techniques are inadequate or too expensive through lack of fossil energy (oil).

Actual introduction of irradiated food, however, needs to be done very carefully. Consequently, supporting research is permanently necessary. The speed at which the procedure comes onto the market depends largely on regulatory policies. Furthermore, information and advice to both producer and consumer are extremely important.

7. REFERENCES

1. Manual of Food Irradiation Dosimetry. Technical Report Series No 178. International Atomic Energy Agency, Vienna, 1977.
2. Tripp, G.E., Packaging for irradiation of foods. International Journal of Applied Radiation and Isotopes 6 (1959) 199-206.
3. Becking, J.H., Radio-sterilization of nutrient media. Miscellaneous Papers 9 (1971) 55-87. Landbouwhogeschool, Wageningen, Netherlands.
4. Langerak, D.Is., De toepasbaarheid van straling bij bewaring van groenten en fruit. Voedingsmiddelentechnologie 3 (1972) 182-184.
5. Langerak, D.Is. & Hovestad, R., Verbetering van de microbiologische kwaliteit van diepvries zilveruitjes door middel van straling. Technical and Preliminary Research Report No 81. Stichting ITAL, Wageningen, Netherlands, 1978.
6. Langerak, D.Is., The influence of irradiation and packaging on the keeping quality of prepacked cut endive, chicory and onions. Acta Alimentaria 4 (1975) 123-138.
7. Langerak, D.Is. & Damen, G.A.A., Influence of irradiation on the keeping quality of prepacked soup-greens stored at 10 °C. Food Preservation by Irradiation. International Atomic Energy Agency-SM-221/41 (1978) 275-282.
8. Langerak, D.Is., Effect van de combinatie verpakken en bestraling op kwaliteit van gesneden groenten. Voedingsmiddelentechnologie 13 (1980) (23) 13-18.
9. Langerak, D.Is. & Bruurs, M.F.J., Preliminary study concerning the influence of combined heat and radiation treatment on the quality of some horticultural products. Acta Alimentaria 2 (1973) 229-243.
10. Brodrick, H.T. & van der Linden, H.J., Radiation preservation of subtropical fruits in South Africa. Atomic Energy Board, Private Bag X 256, Pretoria.

Industrial application of food irradiation

J.G. Leemhorst

1. INTRODUCTION

For more than twenty years now industry has been making large-scale use of the microbicidal effect of radiation. Millions of medical appliances are being sterilized by radiation each year, examples being:
- materials for use in the operating theatre
- dialyse filters
- surgical sutures
- protheses
- hypodermic syringes
- nipples for incubator babies

The two most important sources of industrial radiation are:
- The electron accelerator
- The isotope, Cobalt-60

The limited penetration of the electrons restricts their use; electron radiation for the inactivation of microorganisms is, therefore, only used on a small scale. However, the electron accelerator is being employed to an increasing degree for the processing of plastics. In the past twenty years, gamma irradiation has clearly proved its worth as a sterilization method and is being practised on a large scale.

I shall, therefore, limit myself to discussing the use of gamma irradiation and will take advantage of the experience gained with Gammaster in Ede, the Netherlands.

The introduction of thermoplastic raw materials for the manufacture of disposable medical products in the 1950s created the need for a "cold" sterilization method. Irradation

Food Irradiation Now — ISBN 90-247-2703-0
Copyright Gammaster and Martinus Nijhoff/Dr W. Junk Publishers

proved to be very satisfactory and also offered advantages
over a heat treatment or a "cold" treatment with toxic
gases. A few examples are:
- the temperature rise is negligible
- the sterilant leaves now residue
- the sterilant forms no chemical combination with the product
- the product can be packed in an air-tight condition
- there is only one parameter known to the process, viz. time
- the process is energy-saving

The first gamma irradiation facilities were commissioned
in 1959 in England and France. Since then the number of plants
has steadily increased and the growing resistance towards the
use of toxic gases has given an extra impulse to the industry
during the past few years. In the USA, especially, the number
of facilities has rapidly increased. This concerns plants
that, in American style, have three to six times the production
capacity of those built before 1978.

Irradiation as a sterilization method for medical products
is accepted all over the world (1).
The efficient inactivation of microorganisms and excellent
reproducibility have led the authorities towards a more relaxed regulatory policy. Usually, after industrial sterilization,
random samples are required to prove that the process has been
effective. By gamma irradiation and the generally applied
dose of 25 kGy (1 Gy = 100 rad), inactivation of microorganisms
is so efficient that, for normally occurring bacteria, the
killing efficiency could reach 10^{25} or more (2).

The following is used as the definition of sterility by
gamma irradiation: "The occurrence of not more than one
bacterium per 1 000 000 units". The efficient inactivation
of microorganisms is guarantee that the product, after treatment,
satisfies this definition.

The sterility control of samples is, therefore, of very
little use and almost all countries have accepted the so-called
"dosimetric release". This is a procedure which, by dosimeters,
shows that the correct dose has been received throughout
the product. One dosimeter that is frequently used is a small

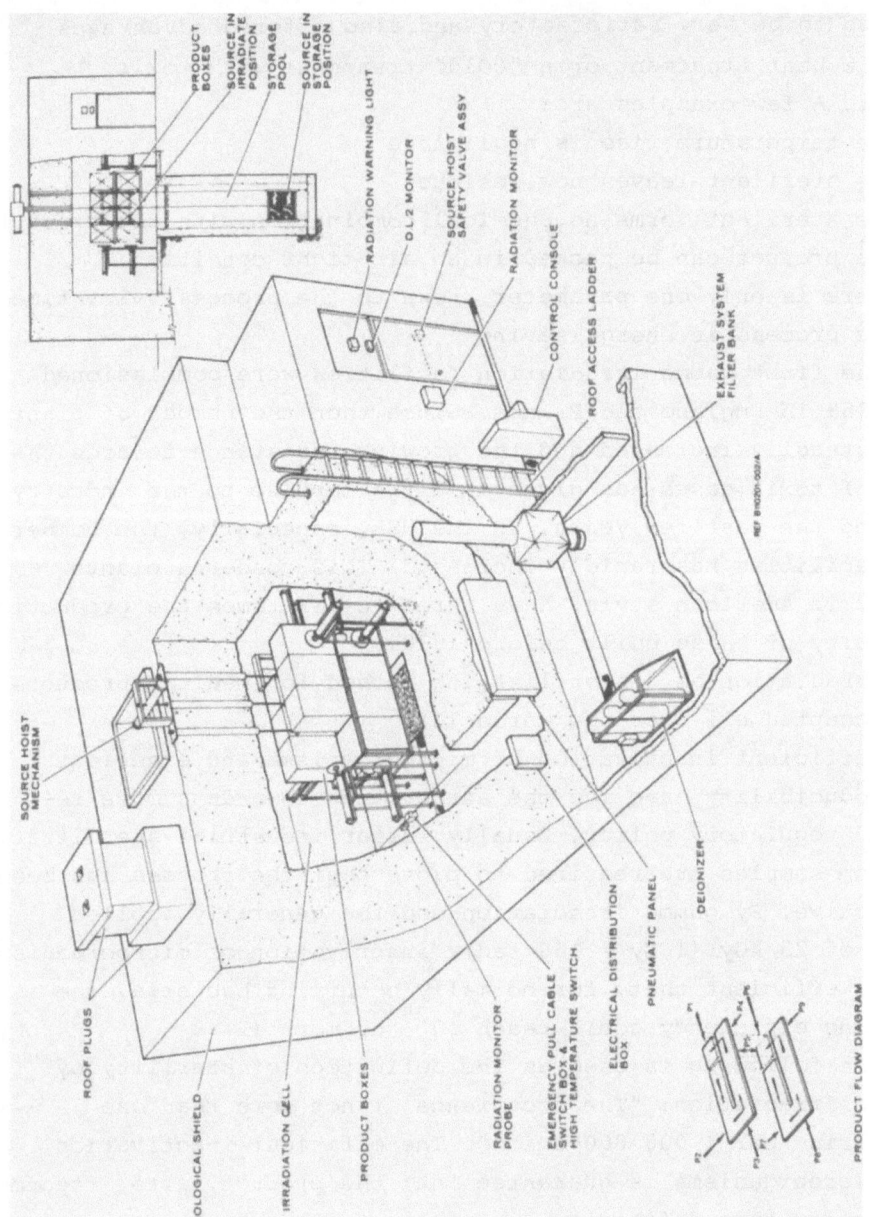

FIG. 1. JS 7200-irradiator.

perspex sheet which changes colour when irradiated (4). The degree of colouring determines the dose received.

Almost all the large manufacturers of medical appliances have their own irradiaton facilities. Others make use of service plants. There is an international agreement concerning the regulations applying to irradiation plants. The process, the process controls, the licencing and registration of the product take place on a "Code of Practice" basis (3).

It is obvious that much of the experience gained in the irradiation of medical appliances is directly applicable to the irradiation of food.

In the past three years we have been irradiating foodstuffs with the Gammaster facility which was originally designed for the sterilization of medical equipment.

A great diversity of products have been irradiated. In spite of some limitations of the facility, the process has proved to be very satisfactory. The technology for medical sterilization is directly applicable. At present, besides the sterilization of medical equipment, an average of twenty tonnes of foodstuffs, mainly spices, grains, herbs and fish products, are being irradiated every week. The Pilot Plant for Food Irradiation handles a similar quantity.

2. PLANT (Fig. 1)

In principle, an irradiation plant is simple in construction. The heart is formed by the radiation source, usually consisting of the isotope Cobalt-60. For specific applications, however, an electron accelerator can be used.

The product to be irradiated passes the source by means of an automatic transport system. The transport route is chosen so that the radiaton penetrates the product from all sides, thus ensuring that it receives a uniform amount of radiation (dose).

The transport system can be a conveyor-belt or a monorail system. Within the labyrinth and irradiation room the product placed in the container or carrier is pneumatically propelled.

The irradiation chamber is surrounded by concrete shielding.

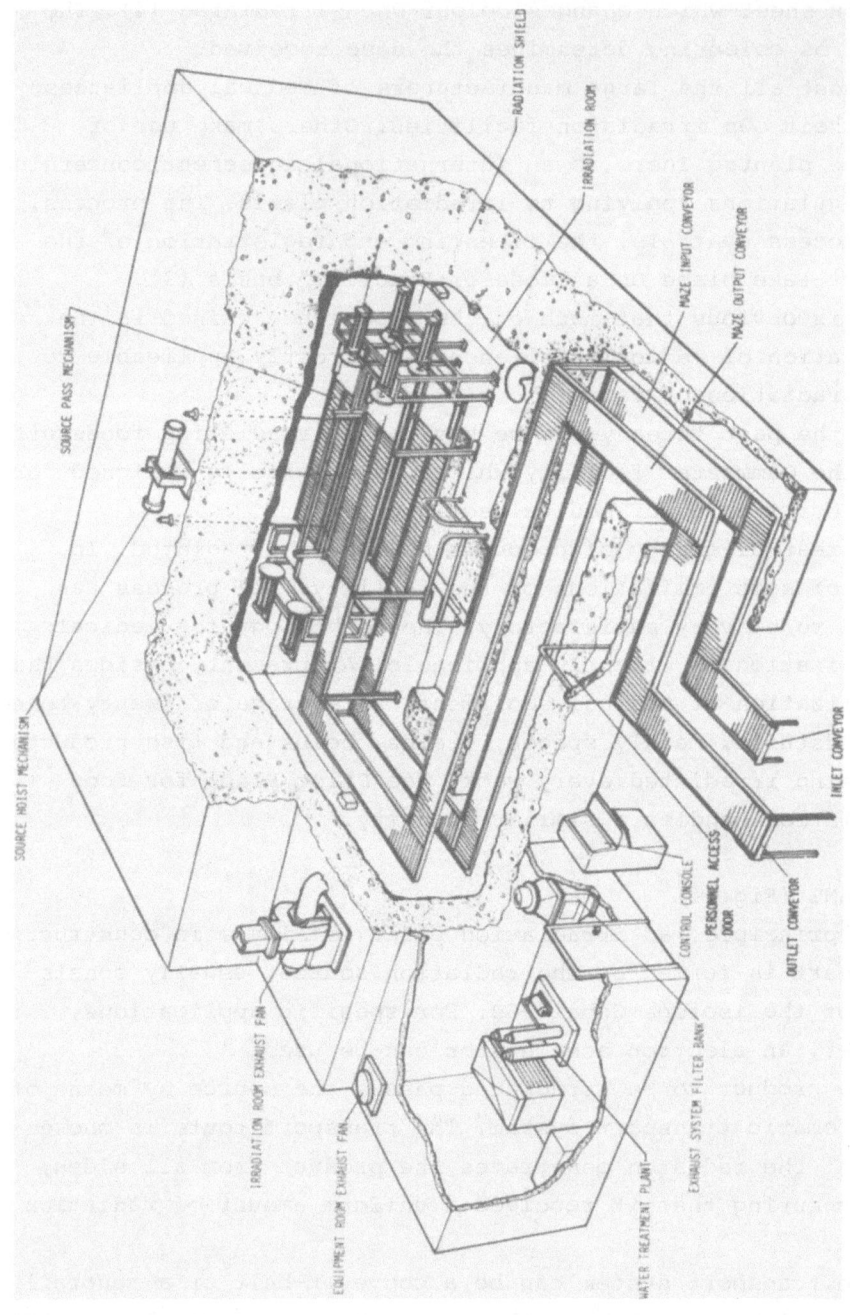

64

FIG. 2. JS 8500-irradiator.

RADIATION SHIELD

IRRADIATION ROOM

MAZE INPUT CONVEYOR

MAZE OUTPUT CONVEYOR

SOURCE PASS MECHANISM

SOURCE HOIST MECHANISM

INLET CONVEYOR

OUTLET CONVEYOR

CONTROL CONSOLE

PERSONNEL ACCESS
DOOR

IRRADIATION ROOM EXHAUST FAN

EQUIPMENT ROOM EXHAUST FAN

WATER TREATMENT PLANT

EXHAUST SYSTEM FILTER BANK

Outside this shielding no radiation is present. The radiation
source can be placed in two positions:
- the irradiation position, around which the products pass;
- the storage position, usually formed by a water pool.

All maintenance and warning systems are integrated in such
a way that for every break in the normal operation the source
is directly lowered into the safe storage position. Entry to
the irradiation chamber can only be gained after a certain
procedure has been followed which gives absolute certainty
that there is no longer any radiation in the chamber.

3. CONTAINER PLANT (Fig. 2)

The Cobalt-60 radiation source gives off continuous energy.
This means that to obtain optimal utilization of the facility,
the transport operation must also be continuous. Many facilities
achieve this by using a large conveyor-belt system, divided
into input and output conveyors.

During a normal eight-hour working day, approximately 300
containers with products are prepared and placed on the input
conveyors. Even during the night, these containers are trans-
ported past the source and sterilized. The following morning
they are emptied and the products despatched.

If there is a breakdown in the operation, the source is
immediately returned to the storage position and the technician
on duty is automatically called.

4. PALLET FACILITY (Fig. 3)

Interest in the irradiation process is so great that the
Gammaster facility is in continuous operation. The construction
of a second plant has, therefore, been started. This second
facility, which is expected to be operational by May/June 1982,
is being built to a completely new design and will not have
the restrictions of the present plant.

The existing facility uses containers, which means that the
packaging of the product must be adjusted to this method. Loading
and emptying of containers is also very labour-intensive. The
transport system is slow, making it impossible to give low

66

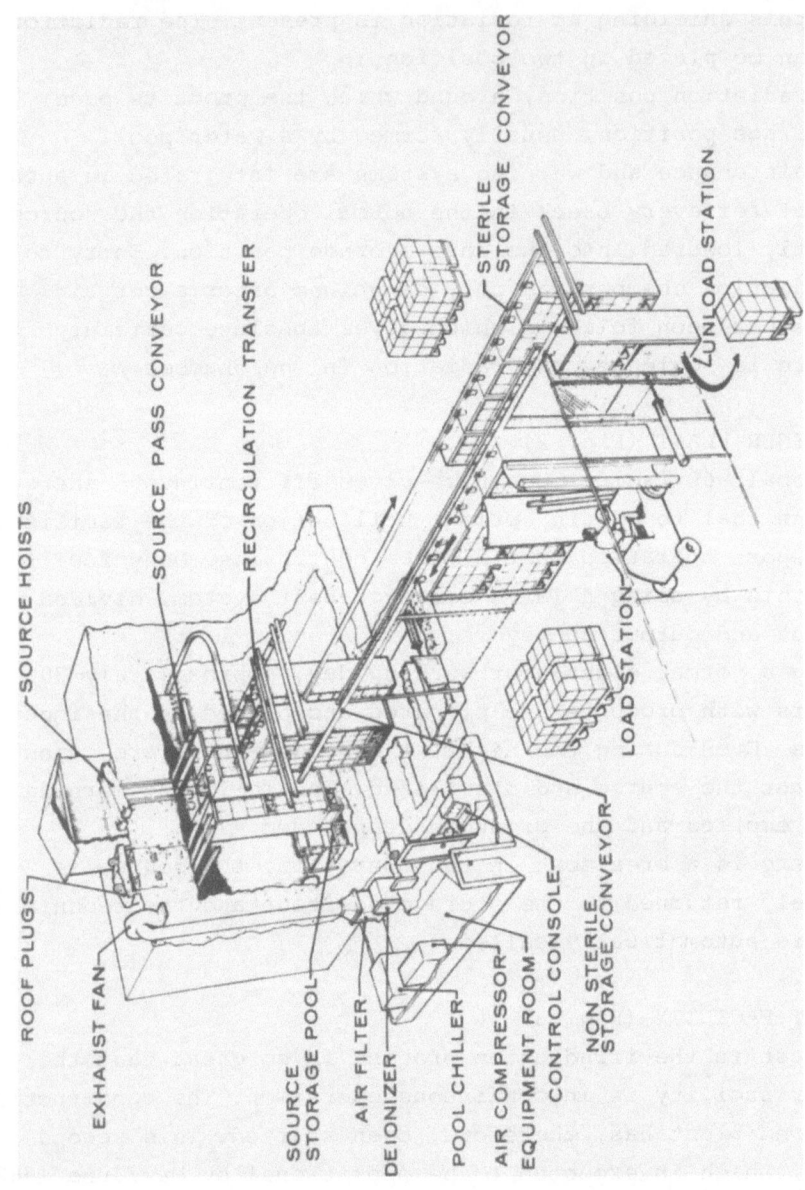

FIG. 3. JS 9000 pallet irradiator.

doses. Approximately 3.5 kGy is the lowest limit.

With the so-called multi-purpose pallet facility these difficulties will be eliminated. With this facility it will be possible to uniformly irradiate the product which will be placed on pallets. Two pallets will be placed in a cage for this purpose. The cage goes through the irradiation chamber in such a way, that a four-sided irradiation takes place, giving very good uniformity. The number of rotations round the source can be programmed for each cage, so that the final dose is a multiple of the chosen dose per circuit. The maximum product measurements are 100 cm x 120 cm x 180 cm. Each pallet can support a weight of at least 1.6 tonnes.

5. PLANT ECONOMY

The energy consumption of the process is low. A comparison between boiling, cooling, freezing and irradiating with 3.0 kGy shows that even in the most unfavourable situation, the energy consumption for irradiation is much less than for the other physical processes (5, 6).

This energy is mainly used to transport the product past the source. With a large source and a fast-moving conveyor, the energy consumption per kilogram drops appreciably.

The costs associated with an irradiation plant are mainly determined by the high initial investment. For a medium-sized plant with shielding, warehouse and radiation source this investment would be somewhere in the region of 1.5 to 3 million dollars. The variable costs are determined by the amount of time the product needs to be effectively irradiated, i.e. the received dose and the costs of maintaining the strength of the source. To be able to run the process economically, it is essential that large volumes are processed and that the plant is working continuously.

6. PRICES

Comparison with other process costs shows that the irradiation process is definitely economically feasible. For example, the costs of irradiation are:

68

Dfl. 0.04/kg for sprout inhibition of onions (0.10-0.15 kGy),
Dfl. 0.09/kg for desinfestation of cereals (0.5-1.5 kGy),
Dfl. 0.15/kg for decontamination of poultry (3.0-4.0 kGy), and
Dfl. 0.30/kg for decontamination of spices (8.0 kGy).

7. CONCLUSION

The unique regulatory policy of the Dutch authorities, combined with the activities of the Pilot Plant for Food Irradiation and Gammaster, has meant that the Netherlands has played a leading role in this field.

People from all over the world come to the Netherlands to follow courses on food irradiation technology. Food irradiation projects have been set up in many countries.

Plants are being built in the USA, South Africa, West Germany, Israel and Hungary.

We expect that, in a very short time, the irradiation of foodstuffs will be taking place on a much larger scale.

8. REFERENCES

1. Leemhorst, J.G., Industrielle Anwendung der Strahlen-sterilisation mit Gammastrahlen. Concept Seminar, Frankfurt, 30 September 1980.
2. Brynjolfsson, A., Food irradiation in the United States. Lecture presented at the International Meat Research Congress, Colorado Springs.
3. Code of Practice. Association Internationale d'Irradiation Industrielle.
4. Chadwick, K.H., Facility calibration, the commissioning of a process and routine monitoring practices. International Atomic Energy Agency-SM-192/30.
5. Brynjolfsson, A., Food; energy; developing countries; food irradiation. International Atomic Energy Agency-SM-250/26.
6. Fraser, F.M. , Gamma Radiation Processing equipment and associated energy requirements. International Atomic Energy Agency-SM-250/4.

The government and food irradiation − National and international rules and regulations

J.Ch. Cornelis

It has always been the task of the Government to protect the collective interests and bona fide trading of the people. It fulfils this task by creating statutory regulations.

The difference between statutory and other types of regulations is that the statutory ones apply to everybody, and their observance can be enforced. Moreover, in some cases, the Government can impose sanctions if they are not observed.

This Government protection can be found in the oldest known legislation and also in the social system of all primitive societies that have no written laws.

The nature and administration of the regulations have varied greatly over the last four thousand years or so, and even now there are still differences between the various communities.

The task of the Government has broadened extensively, especially during the twentieth century. However, the intention of this paper is not to analyse this fact. Just stating it will suffice.

For many years now it has been the responsability of the Government in the Netherlands to ensure that the food eaten by the people is not harmful to health.

Even in the Middle Ages there were statutory regulations, called bye-laws, to protect important matters such as the quality of the beer! These bye-laws only applied to a small area; a town, or a district.

A statute applying to the whole country, the Warenwet (Pure Food Act), was introduced in 1919 and, after many additions and alterations, is still in operation today. One of the main aims of this Act is the promotion of good-quality food. It

Food Irradiation Now − ISBN 90-247-2703-0
Copyright Gammaster and Martinus Nijhoff/Dr W. Junk Publishers

does not, however, apply to meat and meat products. These foodstuffs come under the <u>Vleeskeuringswet</u> (Meat Control Act) of 1919, the aim of which is to exclude meat and meat products that could be harmful to health. There are other acts whose aims are to protect the people against suspect food and drinks, such as the <u>Drankwet</u> (Licencing Law) and the <u>Destructiewet</u> (Law Purporting the Disposal of Carrions), but the Pure Food Act and Meat Control Act are the most important.

It has, therefore, long been the duty of the Government to ensure that foodstuffs are not harmful to health. The history of protection against radiation is much shorter. Ionizing radiation, which has been around since the beginning of time, was only discovered at the end of the past century by Röntgen and Becquerel. X-ray machines came into use straight away. The application of radioactive material remained limited to radium, a relatively easily isolated radioactive element. Ionizing radiation was mainly used for medical purposes.

The use of radioisotopes for various other means only increased after it had become possible to produce them on a large scale. Because ionizing radiation forms a potential hazard for both public health and the environment, it was considered to be the Government's task to guard against these dangers.

After considerable emergency legislation, mainly based on the Pure Food Act and the Nuisance Act in 1957-1970, the <u>Kernenergiewet</u> (Nuclear Energy Act) came into force on 1 January 1970. A main objective of this Act is the "protection of man, animals, plants and goods against the hazards connected with the exposure to radiation".

The subject of this symposium is the application of ionizing radiation to foodstuffs. From the above, it is self-evident that the irradiation of foodstuffs is controlled by two acts: the Pure Food Act and the Nuclear Energy Act.

The Pure Food Act is concerned with the promotion of good-quality food and the Nuclear Energy Act with the protection of man and property against the hazards created by ionizing radiation. Both acts contain ample provisions for reaching the desired objective.

reaching the desired objective.

The Pure Food Act demands certain requirements concerning
the composition and hygiene of foodstuffs. The labelling
(description) of certain products also comes under this act.

The Nuclear Power Act has a licencing system for the application of radioactive materials (Article 29) and the permit may
contain a provision for the protection of man, animals, plants
and goods. Dose determination (dosimetry) is also done on
the basis of the Nuclear Energy Act.

Both these Acts are applicable to food irradiation. Broadly
speaking: the irradiation of foodstuffs is covered by the
Nuclear Energy Act, the irradiated food by the Pure Food Act.

At present the situation is as follows. Because of the
lack of statutory regulations specifically concerned with
irradiated foodstuffs, use is made of the Nuclear Energy Act.
The irradiation process requires a powerful radiation source.
This source can be either a radioactive element, for example
Cobalt-60, or ionizing radiation produced by an electron
generator.

Before such a source can be used, a licence is required,
as laid down by Article 29 or 34 of the Nuclear Energy Act.
Several conditions are attached to this permit. These conditions
are partly the same as those laid down for similar sources
used for other purposes. They concern shielding, safe working
regulations and the prevention of the spread of radioactivity
and radiation in the environment.

For a source intended for food irradiation, extra conditions
are added. Amongst other things, these may concern the dosimetry and the implementation of dose measurements and their
registration. The basis of these extra conditions is, however,
formed by the following. The licence holder, before he starts
to irradiate a certain food product, must obtain special permission from the Minister for Public Health and the Environment. Because food is also the concern of the Minister of
Agriculture and Fisheries, the Minister for Public Health and
the Environment can only give the required permission after
consulting his colleague in the Ministry of Agriculture and

Fisheries.

When requesting permission, the applicant must state why
he is of the opinion that irradiation is not detrimental to
public health. These data cover a broad area, radiological,
bacteriological, chemical, organoleptical, etc. All these
aspects are collected in one word - "wholesomeness". The
data concerning wholesomeness need not necessarily have been
collected as a result of the applicant's own research. This
would be unreasonable. In principle, all data obtained through
reliable research, wherever this was done, are acceptable. The
data are evaluated by a Committee of the Public Health Council.
The procedure of this Committee is described in the lecture
given by R.M. Ulmann.

If the evaluation results in positive advice being given to
the Ministers, then permission to irradiate the foodstuff
concerned with a certain dose is usually granted. It is sometimes necessary to apply still more conditions. For example,
when irradiating frozen products, the temperature may not rise
above a certain value. These conditions are related to the more
general food hygiene regulations. Permission can be granted
in three categories.

Category I states that the product may be irradiated but
not sold to the general public. It may only be consumed for
example by members of a tasting panel.

Category II states that the amount of the product to be
irradiated is limited. This may also mean that the irradiated
product may only be sold to the public through certain indicated
channels, with the aim of investigating public reaction and
keeping quality, etc.

Category III states: unlimited amounts for an unlimited time.

At present, the Ministry for Public Health and the Environment is working, in cooperation with the Ministry of Agriculture
and Fisheries, on a statutory regulation aimed specifically at
the irradiation of foodstuffs. This statute will be based on both
the Pure Food Act and the Nuclear Energy Act. The section pertaining to the Pure Food Act will state the rules concerning
irradiated food and the section pertaining to the Nuclear

Energy Act the irradiation process itself.

Apart from all the legal technicalities there is one question
of a more general nature that is always being asked: How
acceptable is it that if a product is unsuitable for consumption
on its own merits, it becomes suitable by irradiation?

On the one hand it can be argued that irradiation - or any
other method - may not be used to make an unsuitable product
suitable for consumption. This rather general question is asked
especially in connection with the irradiation procedure, because
irradiation is an excellent way of eliminating bacterial infections such as Salmonella.

On the face of it, the answer seems simple enough: it is
not acceptable. My personal opinion is that the answer is not
so simple. For many, many years methods have been used to make
non-consumable products into consumable ones. Grain, potatoes
and any number of other products are unsuitable for consumption
in the state in which they are harvested. They are made suitable
by a heating process such as boiling. The question is, therefore, which method is suitable and which is not suitable. The
fact that a certain method can be applied for thousands of
years, can surely not supply the answer. See, for example, the
warnings against the consumption of grilled and roast meat
that have arisen in the last few years because of the carcino-
gens formed by these processes. Grilling and roasting are
methods that have been used for thousands of years!

Again according to my personal opinion, I do not see why
it should not be acceptable to make a product that contains
dangerous amounts of Salmonella fit for consumption by inac-
tivating these microorganisms.

On the other hand, the approach that hygiene is not important
because a radiation treatment takes care of everything should
also not be encouraged.

One may also wonder why so much emphasis needs to be placed
on the irradiation facility. The reason is that radiation,
when applied at the correct dose, leaves no trace and,
therefore, it is impossible to prove whether or not a product
has been irradiated let alone with which dose, leaving the

radiation facility as the only centre of control.

So far, the legal aspects have only been discussed from a national point of view. The international aspect, however, must also be examined. Food marketing is, after all, of international concern.

There are certain factors that play a role when trying to reach agreement on international regulations.

The simplest one is the reaching of an international agreement on the maximum permissible dose. These limits are the result of years of international research carried out by, amongst others, the International Project in the Field of Food Irradiation.

The problems can be summarized as follows.
- It is the task of national authorities to make sure that the foodstuffs consumed in their own country are not damaging to public health.
- Where foodstuffs that have been treated by one or other preservation process in their own country are concerned, the authorities are able to follow the whole process; if, however, foodstuffs that have been treated in a different country are concerned, the authorities are not in the position to judge the whole process.
- Where, for example, the addition of a chemical preservative is concerned, the Government of the importing country is not powerless; analysis of the imported product will show which, and how much, preservative has been used.
- In order to make international marketing of foodstuffs possible, it is necessary, in the case of chemical preservatives, to reach international agreement and to lay down international regulations concerning chemical additives that may or may not be used, and the limits of their application. Each Government can independently determine whether a certain consignment of imported foodstuffs satisfies the international regulations.
- With the irradiation procedure this is not possible. The irradiated product is not changed by the irradiation process.
- This means that international agreements must be more detailed than is necessary in other cases. Not only is agreement necessary

about the way in which the procedure may be applied, including dose limits, etc., but a means of indicating that a certain product has in fact been irradiated must be agreed upon and, above all, this means must offer sufficient guarantees to the national authorities that the procedure has been applied in the agreed manner.

A solution to this problem is well on the way. Within the framework of the Codex Alimentarius two "Standards" concerning the application of radiation are in the final stages of discussion. If these Standards are accepted on an international basis it will be a very important step. The enforcement of the Standards only solves half the problem. Much more is needed. This was stipulated in a report of a joint WHO/FAO/IAEA Advisory Group, who met in Wageningen in 1977, entitled: International Acceptance of Irradiated Food.

The following is taken from the conclusions of this Panel.

The intention is not to set aside the still debated international regulations that appear in connection with the Codex Alimentarius. The intention is to arrange for additional regulations, which supplement the Codex Standards in areas not covered by the Codex organization.

I have tried personally to help in finding a solution, and some of my suggestions are set down in a paper for the Analysis of the International Legal and Administrative Regulations on Food Irradiation with Regard to the Public Health Aspects.

International discussions have been waiting for the report on Wholesomeness of Irradiated Food, which has been discussed elsewhere.

The Board of Management of the International Project in the Field of Food Irradiation (this project has been completed in 1981) has also expressed the wish for these international discussions to get quickly under way.

The same international organizations who convened the Joint Expert Committee, namely the FAO, IAEA and WHO, will shortly be sending a letter to all member states proposing the promotion of further cooperation. It is to be hoped that this proposal finds the necessary response.

REFERENCES

1. Cornelis, J.Ch., Analysis of the international legal and
   administrative regulations on food irradiation with regard
   to public health aspects. EUR 4466c (1970).
2. Report of a consultation group on the legal aspects of food
   irradiation organized by FAO, IAEA, WHO. International
   Atomic Energy Agency, Vienna, 1973, STI/Doc/59.
3. Legal Series No. 11. International acceptance of irradiated
   food. Legal aspects. Report of a joint FAO/IAEA/WHO Advisory
   Group on international acceptance of irradiated food,
   Wageningen 28 November - 1 December 1977. International
   Atomic Energy Agency, Vienna, 1979.
4. Legal, administrative and psychological barriers to the
   industrial application of food irradiation and the trade
   in irradiated food. In: J.Ch. Cornelis, Food preservation
   by irradiation, Vol. 2. International Atomic Energy Agency,
   Vienna, 1978.

14 Years clearing irradiated foods in the Netherlands

R.M. Ulmann

Around 1900 the destroying capacity of X-rays on pathogenic microorganisms was discovered. This discovery was patented some years later. Consequently, food irradiation was - scientific-ally - born.

The practical application had to await the technological development and feasability of ionizing radiation, as is the case with so many new scientific discoveries. When nuclear fuel plants provided gamma-rays as a "waste"product the practical use began. In the mid-fifties, the United States of America started a large research programme: the Army (Natick) studying radappertization, the Atomic Energy Commission studying radurization.

Not long afterwards the Netherlands and specifically ITAL (Institute for the application of atomic energy in agriculture) in Wageningen, started to investigate the application for food preservation. Positive and optimistic results from the UK, USSR, USA and France led to the decision to construct a Food Irradiation Pilot Plant in the Netherlands in 1966. Before the end of 1967, testing had begun at this facility.

Meanwhile, France practised food irradiation using a mobile source called IRMA (IRradiations Mobile Autonome, Fig. 1). It was not only used in France but also in interested neighbouring countries, visiting Wageningen in 1966. This made our health authorities aware of the necessity to produce legislation for the clearance of irradiated foods.

Authorization may arise from two principles: one prohibiting everything that is not permitted, the other permitting everything that is not strictly forbidden. The former is not very

Food Irradiation Now – ISBN 90-247-2703-0
Copyright Gammaster and Martinus Nijhoff/Dr W. Junk Publishers

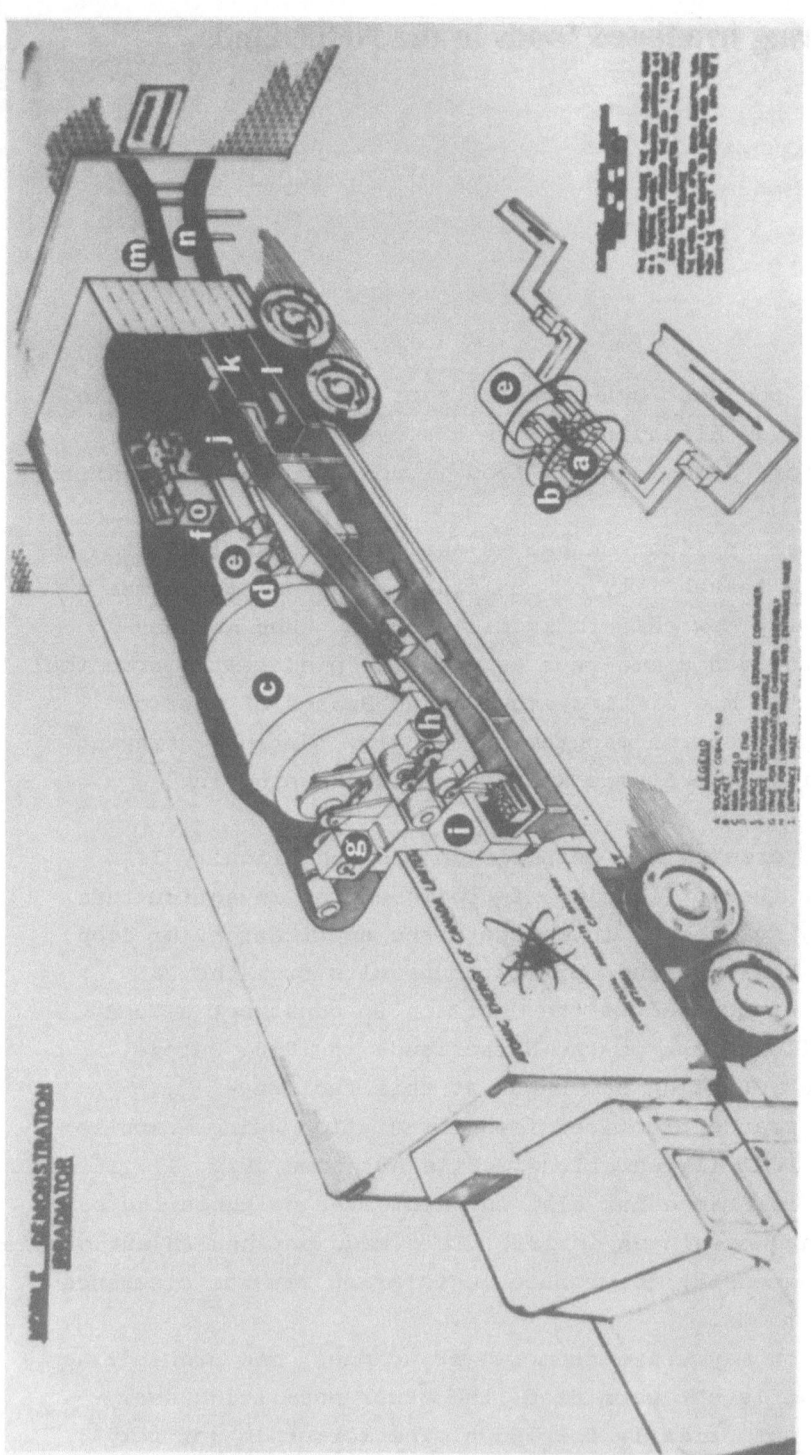

Fig. 1. IRMA Mobile demonstration irradiator. Legend: A, source-cobalt 60; B, bucket; C, main shield; D, removable end; E, source mechanism and storage container; F, source positioning handle; G, drive for irradiation chamber assembly; H, drive for loading product into entrance maze; I, entrance maze; J, drive for discharging product onto exit conveyor; K, powered entrance conveyor; L, powered exit conveyor; M, portable powered loading conveyor; N, portable powered discharge conveyor; O, control console.

practical; especially as far as irradiated foods are concerned. It was soon obvious that irradiation causes so few alterations in the food that it proved almost impossible to detect and prove food as irradiated. Hence this principle of repressive control was not introduced. Instead, a positive clearing policy was introduced, gradually permitting - groups of - products after thorough investigations.

This work was performed by the scrutinizing body, whose creation has not yet been described. On 1 August1967, Dr. Kruisinga - at that time Under-secretary of Health - invited the chairman of the Health Council to form a body "whose task it should be to authorize the Pilot Plant for Food Irradiation to irradiate products under certain conditions after careful research". How dynamic and progressive he was, was proved in the same letter stating: "It is intended to consult this body at other events such as the forthcoming visit of the mobile Euratom irradiation unit IRMA." As can be seen from this context, a subsidiary concern of Euratom, Eurisotop, had meanwhile taken over IRMA from the French in order to promote it within the European community.

At that time the future work of the scrutinizing body was defined and could start after members had been invited and installed. Its task and scope was not restricted to one - irradiation - plant. Although expert members were personally invited, they were recruted from official organizations: Ministry of Health, Ministry of Agriculture, Food Inspection, Feed Inspection, Nutritional Council and several others. The scrutinizing body was soon nicknamed "Watchdog Committee" because of its work in clearing petitions. The same under-secretary decided shortly afterwards that the scrutinizing body should only advise him, and that he would authorize.

The main parameters stated when filing petitions were microbiological, nutritional and toxicological data. As a two-year programme for chronic toxicological research was too costly, many foreign results were used in the beginning.

Clearances (Table 1) were given in three categories:
I: not for trade purposes, maximum quantity 1.000 kg;

Table 1. Clearances issued in the Netherlands (status, June 1981).

Product	Purpose of irradiation	Max. dose (kGy)
Cocoa beans	disinfestation	0.7
Strawberries	radurization	2.5
Asparagus	radurization	2.0
Shrimps	radurization	1.0
Spices and condiments	radicidation	10.0
Poultry	radurization	3.0
Potatoes	sprout inhibition	0.15
Onions	sprout inhibition	0.05
Batter mix	radicidation	1.5
Prepacked, peeled potatoes	radurization	0.5
Fish fillets	radurization	1.0
Frog legs (deep frozen)	radicidation	5.0
Rice	disinfestation	1.0
Shrimps (deep frozen)	radicidation	7.0
Rye Bread	radurization	5.0
Mushrooms	growth inhibition	2.5
Food packaging materials		
Deep frozen meals[1]	radappertization	25.0
Fresh, tinned and liquid foodstuffs[1]	radappertization	25.0
Milk blanks		10.0
Prepacked, cut endive	radurization	1.0

1. For hospital patients in reversed barrier isolation.

II: restricted quantities, mostly a few tonnes, destined
   for test marketing;

III: unrestricted quantities with an unconditional clearing
    (ref. Cornelis).

There was a great feeling of satisfaction when the first
number III clearance could be granted for mushrooms in June
1970. Most of the required data stemmed from Dutch research.
Unfortunately, the marketing of this product was very disap-
pointing. This was caused by circumstances beyond the scope
of either the scrutinizing body or the Pilot Plant for Food
Irradiation. The Dutch eventually lost their lead. They have,
however, caught up again during recent years thanks to the
new irradiation plant in Ede.

The positive recommendations of JECFI - the WHO/FAO Joint
Expert Committee on Food Irradiation - in 1975 and especially
in 1980 helped many a hesitating industry to become convinced
of the feasibility and safety of the irradiation process.

The work within the scrutinizing body was distributed in
such a way that everybody judged from within his own discipline.
Internally, this often led to severe disagreement on the
scientific interpretation or the practical feasibility of
the reported data, but the final advice was always unanimous.
Sometimes, a clearance as the result of the produced scientific
data had practical consequences on our national export.

Strictly speaking this was not the task nor was it within
the terms of reference of a Health Council Committee, but it
was of paramount importance for the effectuation: how will
neighbouring countries react to a national clearance? Would
its effect be negative press releases such as "Bestrahlte
Hühner aus Holland" (irradiated chicken from Holland) or would
the trade frontiers be closed on unfounded grounds as was
the case with the United Kingdom, which forbade Dutch tomatoes
because of an outbreak of swine fever? That these health
barriers do not only exist in our imagination has been proved
by recent events concerning Dutch shrimps.

The import of irradiated products is a similar problem.
As already explained, repressive control is difficult to

realize. It is much better, according to the Dutch philosophy,
to permit the import provided the irradiation is authorized
by the exporting country and that the conditions concerning
that authorization are comparable with ours. This is already
the case with Israeli strawberries and South-African mangoes.
These two examples prove that we work on scientific and not
on political grounds!

The report and recommendations of the Joint Expert Committee
on Food Irradiation have given the green light to this technology. Hence, toxicological testing of not yet investigated
products has become superfluous. Consequently, the scrutinizing
body will no longer be necessary when the national Codex
Alimentarius has legal force. Its work could be discontinued,
its members dismissed. Yet it does not seem a wise policy.

Firstly, the greater part of food irradiation has been
cleared (desinfestation and radurization) but not yet all
(radicidation and radappertization). Secondly, it would be
a waste of effort and expertise to have all their mutual
knowledge dispersed. For instance, apart from foodstuffs
there are also pharmaceutical products to be treated, and
the knowledge gained from the former could be used for the
latter. Thirdly, there is no reason to take such a decision
before an international agreement has been decided upon.

Should our independent scrutinizing body be transformed
into a civil service committee? This also does not seem a
wise decision since its complete independency has often
promoted a swift decision policy. There is another Health
Council committee on which ours could be modelled in the
future.

I should like to conclude my address with a salute to the
past and a look into the future. The salute concerns the
members of the scrutinizing body. One feeling is paramount:
the will to agree in the interest of public health has always
prevailed over personal opinion and interests. The look into
the future concerns our host Gammaster whose hospitality has
gathered us all here today.

May this company carry the banner of technological lead

with the construction and installation of the new low-dose
plant and may the advantages be available to both Dutch and
foreign food industries.

International aspects of food irradiation

J.G. van Kooij

1. INTRODUCTION

Twenty-five years of research and development work on technological aspects of food irradiation have clearly demonstrated
the practicality of the use of irradiation in the preservation
of foods. Moreover, basic research in radiation biology and
chemistry of food along with negative evidence from animal
feeding studies have demonstrated that food irradiated up to
10 kGy is safe for consumption by man. Because of the great
need for an alternative to additives and fumigants which pose
hazards for consumers as well as for workers in food processing
factories, the possibilities of the irradiation technology to
replace or drastically reduce the use of such chemical preservatives are now being investigated in many countries. Food
irradiation is energy-conserving when compared with conventional
large-scale used methods of preserving food to obtain a similar
shelf-life. Analysis of the economics of food irradiation
processing has made it clear that successful and profitable use
of this technology requires commercially organized management
practices. The introduction of new technologies is difficult
in many countries due to their economic status. Several developing countries often strive with limited technical know-how
and with inadequate infrastructure to adopt food irradiation
efficiently. Furthermore, full commercialization of the food
irradiation process is still tempered by the lack of a general
acceptance in many countries. Improvement of this situation is
largely beyond control of private business and requires examination and resolution at the international level.

    Achievement of a general acceptance of food irradiation

Food Irradiation Now – ISBN 90-247-2703-0
Copyright Gammaster and Martinus Nijhoff/Dr W. Junk Publishers

depends on the following three major factors:
- clearance of the process in many countries;
- acceptance of the Standard for Food Irradiation;
- harmonization of legislative requirements regarding the control of the food irradiation process and of the international trade in irradiated foods.

Important developments in the field of clearance, standardization, and legislation and their present status are the subjects of this paper.

2. TOWARDS ACHIEVING GENERAL ACCEPTANCE OF FOOD IRRADIATION

During the past decade world-wide interest has been shown in the aspects of clearance, standardization, and legislation of food irradiation by several international organizations and by public health authorities in various countries. Table 1 summarizes activities which were initiated by the Joint FAO/IAEA Division or developed in close cooperation with the World Health Organization (WHO), the Nuclear Energy Agency (NEA) of the Organization for Economic Cooperation and Development (OECD), and the Joint FAO/WHO Food Standards Programme of the Codex Alimentarius Commission (CAC) with the purpose to achieve general acceptance of the food irradiation process.

For detailed information on the various activities mentioned in Table 1 readers are referred to the references listed in the same table.

3. EVALUATION OF THE WHOLESOMENESS OF IRRADIATED FOOD

In the sixties national public health authorities displayed a very cautious attitude towards the approval of irradiated food for general sale. This is reflected in the limited number of clearances granted up to 1969, and then only in a few countries (cf. Table 2). The various Joint FAO/IAEA/WHO Expert Committees on the Wholesomeness of Irradiated Food (JECFI) have played a decisive role in the development of an international consensus of the wholesomeness of irradiated food. The 1969 JECFI contributed much to the scope of scientific studies then considered appropriate to establish safety of irradiated food. This and the recognition that more clearances in more

Table 1. International activities between 1969 - 1981.

Joint FAO/IAEA/WHO Expert Committee on the Wholesomeness of
  Irradiated Food (JECFI), 1969 (Ref. 1)
Establishment of the International Project in the Field of
  Food Irradiation (IFIP), 1970
FAO/IAEA Consultants Meeting on Microbial Aspects of Food
  Irradiation, 1974 (Ref. 2)
Joint FAO/IAEA/WHO Expert Committee on the Wholesomeness of
  Irradiated Food, 1976 (Ref. 3)
FAO/IAEA Advisory Group on Standardization of Food Irradiation,
  1976
Codex Committee on Food Additives (CXFA), 1977 (Ref. 4)
FAO/IAEA/WHO Advisory Group on International Acceptance of
  Irradiated Food, 1977 (Ref. 5)
Codex Alimentarius Commission (CAC), 1978 (Ref. 6)
Codex Committee on Food Additives, 1978 (Ref. 7)
Publication of Model Regulations for Control of and Trade in
  Irradiated Food, 1979 (Annex 7 of Ref. 5)
Codex Committee on Food Additives, 1979 (Ref. 8)
Codex Alimentarius Commission, 1979 (Ref. 9)
Joint FAO/IAEA/WHO Expert Committee on the Wholesomeness of
  Irradiated Food, 1980 (Ref. 10)
Codex Committee on Food Additives, 1980 (Ref. 11)
Joint FAO/IAEA/WHO Consultation Group on Revision of the
  General Standard and Code of Practice of Irradiated Foods,
  1981 (Ref. 12)
Codex Alimentarius Commission, 1981 (Ref. 13)
FAO/IAEA Regional Seminar Programme 1981-1983

---

countries would depend on demonstrating scientific evidence
of the wholesomeness of more irradiated foods obtained from
studies supported by many countries, have led to the setting
up of the International Project in the Field of Food Irradiation
(IFIP) in 1970. The Project's major objectives were the wholesomeness testing of irradiated food and research on and inves-

Table 2. Clearances[1] of irradiated foods until 1969.

Product	Country
Deep-frozen meals	Netherlands
Food for hospital patients	United Kingdom
Dry food concentrates	USSR
Culinary prepared meat	USSR
Chicken	USSR
Semi-prepared meat	USSR
Wheat flour and whole wheat flour	Bulgaria, USA
Grain	USSR
Cocoa beans	Netherlands
Strawberry	Netherlands
Asparagus	Netherlands
Mushrooms	Netherlands
Fresh fruit and vegetables	USSR
Dried fruits	USSR
Onions	Canada, Israel, USSR
Potato	Canada, Hungary, Israel, Spain, USA, USSR

1. Not differentiated in categories of clearances

tigations into the methodology of wholesomeness testing. On
the basis of the first studies of IFIP, the 1976 JECFI recommended unconditional acceptance for irradiated wheat and
ground, wheat products, potato, chicken, papaya, and strawberry,
and provisional acceptance for irradiated onion, fresh cod
and redfish, and rice. All these commodities are irradiated
at levels below 10 kGy. Moreover, JECFI also declared the
process of food irradiation a physical process akin to heating
or freezing foods for preservation.

Several countries have reacted positively to the recommendations of the 1976 JECFI as can be seen from Table 3.

88

Table 3. Clearances of irradiated foods since 1976.

Country or organization	Product	Category of acceptance[1]
Argentina	Potatoes	Unconditional
Australia	Frozen shrimps	Provisional
Belgium	Potatoes	Provisional
	Strawberries	"
	Black pepper	"
	Shallots	"
	Garlic	"
	Onions	"
Czechoslovakia	Potatoes	Provisional
	Onions	"
	Mushrooms	"
France	Onions	Provisional
	Garlic	"
	Shallots	"
Hungary	Mixed dry ingredients for canned hashed meat	Provisional
Japan	Onions	Unconditional
Netherlands	Peeled potatoes	Provisional
	Chicken	Unconditional
	Soup greens	Provisional
	Frozen frog's legs	"
	Rice, ground rice products	"
	Rye-bread	"
	Various fish fillets	"

As a result of many further studies and investigations, commissioned and funded by IFIP since 1976, the 1980 JECFI was able to recommend the acceptability of food irradiated up to an overall average dose of 10 kGy. It also concluded that irradiation of any food commodity up to that dose presented no toxicological hazard; hence foods treated in this way no longer need to be tested for toxicity.

Important applications in the low-dose field (e.g. sprout inhibition, insect disinfestation, ripening delay in fruits) and medium-dose field (e.g. reduction of microbial contamination,

Table 3. (continued)

South Africa	Potatoes	Unconditional
	Onions	"
	Garlic	"
	Chicken	"
	Papaya	"
	Mango	"
	Strawberries	"
	Dried bananas	Provisional
	Avocados	"
Joint FAO/IAEA/WHO		
Expert Committee, 1976	Potatoes	Unconditional
	Onions	Provisional
	Papaya	Unconditional
	Strawberries	"
	Wheat and ground wheat products	"
	Rice	Provisional
	Chicken	Unconditional
	Cod and redfish	Provisional

1. The term "provisional" refers to various sorts of acceptance, e.g. experimental batches for pilot studies, marketing tests, limited in time and/or in quantity.

reduction of non-sporing pathogenic microorganisms, improvement of technological properties of food), all fall within this recommended limit.

With the clarification of the wholesomeness of irradiated foods (up to 10 kGy), the door is now open for clearance of more irradiated foods and/or the process of food irradiation by national public health authorities in more countries. In their decisions, national authorities are usually guided by the recommendations or evaluations of international agencies, especially WHO. Public health approvals for irradiation of food and marketing of irradiated food can only be granted by national authorities. At present, the relevant authorities in 23 countries have given unconditional or provisional clearance to 38 items of food and groups of related food products (see Table 4).

Table 4. Year of first clearance in a country for a given food item or group of related products.

PRODUCT \ COUNTRY	AUSTRALIA	BELGIUM	BULGARIA	CANADA	CHILE	CZECHOSLOVAKIA	DENMARK	FRANCE	GERMANY, FED. REP.	HUNGARY	ISRAEL	ITALY	JAPAN	NETHERLANDS	PHILIPPINES	SOUTH AFRICA	SPAIN	THAILAND	UNITED KINGDOM	URUGUAY	U.S.A.	U.S.S.R.
COALFISH, WHITING AND PLAICE FILLETS														76								
RYE BREAD														80								
RICE AND GROUND RICE PRODUCTS														79								
PAPAYA																78						
FROZEN SHRIMPS	79																					
FROZEN FROGLEGS														78								
AVOCADOS																77						
DRIED BANANAS																77						
SOUP GREENS														77								
PEELED POTATOES														76								
MIXED DRY INGREDIENTS FOR CANNED HASHED MEAT									76													
MIXED SPICES									74													
SHALLOTS		80				77																
FRESH, TINNED AND LIQUID FOOD														72								
DEEP-FROZEN MEALS (HOSPITAL)										72				69								
FOOD FOR HOSPITAL PATIENTS																			69			
DRY FOOD CONCENTRATES		72																				66
COD AND HADDOCK FILLETS				73										76								
SHRIMPS														70								
CULINARY PREPARED MEAT																						67
CHICKEN				73										71/76		78						66
SEMI-PREPARED MEAT																						64
WHEAT, FLOUR AND WHOLE WHEAT FLOUR				69																		
WHEAT AND WHEAT FLOUR																					63	
GRAIN		72																				59
SPICES AND CONDIMENTS		80								74				71								
ENDIVES														75								
POWERED BATTER-MIX														74								
VEGETABLE FILLING														74								
COCOA BEANS														69								
MANGOES																78						
STRAWBERRIES		80								73				69		78						
ASPARAGUS														69								
MUSHROOMS							76							69								
FRESH FRUITS AND VEGETABLES		72																				64
DRIED FRUITS		72																				66
GARLIC	80	72				77						73				78						
ONIONS	80	72	65		76	77				73	68	73	80	71/75		78	75	73				67/73
POTATOES	80	72	60	74	76	70	72	74	69	67	73	72	70	72		77	69			70	64	58

4. STANDARDIZATION AND LEGISLATION OF IRRADIATED FOOD

For a world-wide introduction of food irradiation it is necessary to develop national legislation and regulatory procedures that will enhance confidence among trading nations that foods, irradiated in one country and offered for sale in another, have been subjected to commonly acceptable standards of good irradiation practice and control and irradiated in facilities licensed and registered by the competent national authority. To aid harmonization of national laws, the Codex Alimentarius Commission (CAC) has adopted in 1979 a Recommended International General Standard for Irradiated Foods and a Recommended International Code of Practice for the Operation of Radiation Facilities used for the Treatment of Foods.

Both the General Standard and the Code of Practice were developed in accordance with the Codex Procedure for the Elaboration of World-wide Codex Standards by the intergovernmental Codex Committee on Food Additives (CXFA) in close cooperation with the International Atomic Energy Agency. The General Standard took into account the findings of various JECFIs, in particular the recommendations developed by the Expert Committee convened in 1976. The latter committee recommended acceptance of the eight foods mentioned in Table 3 given under "Joint FAO/IAEA/WHO Expert Committee". These recommendations as well as the progress in the clearance for human consumption of certain irradiated foods achieved in various countries, necessitated the development of the General Standard and the Code of Practice for foods processed by irradiation. The first draft of these documents was prepared by the FAO/IAEA Advisory Group on Standardization of Food Irradiation in 1976. The General Standard and the Code of Practice were subject to discussion at the CXFA meetings in 1977, 1978 and 1979, respectively, and adopted by the CAC in 1979. The adopted General Standard and the Code of Practice, both designed to permit free trade among Member States of the CAC, has meanwhile been distributed for government acceptance.

The findings and statements of the 1980 JECFI called for a revision of the General Standard for Irradiated Foods. Revised

drafts of the General Standard and the Code of Practice for
Irradiated Foods were prepared by a Joint FAO/IAEA/WHO Consultation Group which met 1-3 July 1981 in Geneva, Switzerland.
Approval for introduction of the revised documents at Step 3
of the Procedure for the Amendment of Codex Standards was
given by the CAC at its 14th Session which was held from
29 June to 10 July 1981 in Geneva. The revised drafts of
the Standard and the Code of Practice have been distributed
to the CAC Member States for comments. Subsequent review of
these documents and of the comments received from Member States
and from the international organizations concerned has taken
place at the 15th Session of the CXFA, which was held from
16 to 22 March 1982. The major amendments proposed in the
revised draft of the Recommended International General Standard
for Irradiated Foods are the following:
- inclusion of X-ray sources in the list of acceptable
radiation sources;
- stipulation that the overall average dose absorbed by a
food subjected to radiation processing should not exceed
10 kGy;
- a repetition of irradiation within the overall average dose
of 10 kGy should be limited to the case of food commodities
of low moisture content in which re-infestation by insects
cannot be effectively prevented under practical conditions
of storage and transport;
- repetition of irradiation is also considered acceptable when
the food to be irradiated is a processed form of a food that
has already undergone low-dose treatment and when it includes
irradiated minor ingredients;
- omission of the mandatory requirement to label foods for
retail sale that they have been treated with ionizing radiation.
However, the identity of irradiated food up to the point of
retail sale including food handled in bulk and food moving
in international trade is being maintained.

5. HARMONIZATION OF REGULATORY CONTROL OF THE FOOD IRRADIATION
PROCESS

The availability of a Recommended International General
Standard for Irradiated Foods and a growing number of clearances of irradiated foods or of the food irradiation process
in more countries are, of course, promising developments and
important for the commercial introduction of food irradiation
at the national level. International trade in irradiated foods
requires approval for importation by the relevant authorities
of the importing countries. To this effect a Joint FAO/IAEA/WHO
Advisory Group on International Acceptance of Irradiated Food
recommended in 1977 the preparation of Model Regulations for
the Control of and Trade in Irradiated Food. The Model Regulations were published by IAEA in 1979 (Ref. 5). They have been
prepared on the assumption that countries interested in international trade of irradiated foods will accept the Recommended
International General Standard for Irradiated Food and intend
to incorporate the provisions of this Standard into appropriate
national legislation. The Model Regulations were based on the
provisions of the Standard for Irradiated Food adopted by CAC
in 1979. Because of the recommendations of the 1980 JECFI,
consequential amendments to this Standard have meanwhile been
initiated at Step 5 of the Codex Procedure for the Amendment of
Adopted Standards. It is clear that the Model Regulations will
have to be revised in the light of the 1980 JECFI findings and
be based on the provisions of the proposed revised draft
General Standard.

6. PURPOSE OF THE MODEL REGULATIONS

The purpose of the Model Regulations is seen as providing
either:
- the basic for a multilaterial agreement governing the control
of irradiated foods moving in international trade among a
group of countries which have accepted the Standard for
Irradiated Foods and its amendments, or
- a common pattern for individual national legislation in
order to ensure a similar and effective control over the

irradiation of food.

In either case the Model Regulations should be completed by
such supplementary provisions as may be necessary to satisfy
the particular systems of each country provided, however,
that these do not conflict with the purposes and structures
of the Regulations. The term "Model Regulations" must be given
a broad interpretation. In some countries it may constitute
a framework law passed by parliament while in other countries
it could constitute a general regulation implemented by the
executive power.

7. PRESENT STATUS OF ACCEPTANCE OF FOOD IRRADIATION

As can be seen from Table 4 the clearance position
in developing countries is lagging behind the situation
in advanced countries. Food quality control still has to be
established or needs to be reviewed in many developing countries.
In this connection, a Regional Seminar Programme on the
implementation of food irradiation has been initiated for
developing countries with the following objectives:
- to assist, where necessary, national authorities in their
consideration of the acceptance of irradiated food;
- to assess country needs for the development of integrated
food irradiation control systems;
- to assist in the implementation of the Recommended International General Standard for Irradiated Foods and of the amendments
at present being considered by the Codex;
- to focus on the practicality of the use of irradiation for
the preservation of foods.
An enquiry carried out in November 1980 in countries of Asia
and the Pacific indicated that only Thailand has implemented
under the Food Control Act the production, distribution, sale,
import and export of irradiated food which is subject to
licensing by the Food and Drug Administration of the Ministry
of Public Health. The issuance of food irradiation legislation
is being considered in Indonesia, Korea, and Malaysia. No
legislation on food irradiation existed at that time in
Bangladesh, India, Pakistan, the Philippines, Singapore, and

Sri Lanka. For further information on countries' legislation for irradiated foods the readers are referred to Annex 6 of Ref. 5.

Harmonization of national legislation on food irradiation would also be appropriate for many advanced countries in order to aid international trade of irradiated food. In this respect the initiative of the Food Irradiation Working Group of the European Society of Nuclear Methods in Agriculture (ESNA) to organize a workshop on the aspects of the harmonization of food irradiation legislation warrants full support of the participating countries.

Of great importance to the international acceptance of irradiated food is the recently published Federal Register Advance Notice of Proposed Rulemaking for Irradiated Food (Ref. 14). The Food and Drug Administration of the United States of America is considering a change in the criteria for establishing the safety of irradiated foods. Greatly relaxed criteria for approval of the domestic sale of irradiated foods were submitted (Ref. 14) and can be summarized as follows.
- Food irradiated at doses of 1 kGy (100 krad) or less will be considered wholesome and safe for human consumption.
- Food irradiated at doses exceeding 1 kGy will be subject to toxicological testing consisting of a battery of four short-term mutagenicity tests and two 90-day feeding studies (one rodent and one non-rodent mammalian species).
- A food class comprising no more than 0.01% of the daily diet and irradiated at doses of 50 kGy or less will be considered safe for human consumption without toxicological testing.
Considerable concern was expressed (Ref. 15) in the USA regarding the 1 kGy dose limit which is not as broad as the one recommended by the 1980 JECFI. The latter Committee met almost four months after the FDA Working Group sent its recommendations to the Director of the Bureau of Foods. As mentioned in the Advance Notice, the 1980 JECFI report will be considered in the development of further actions governing the aspects of clearance and legislation of irradiated foods.

8. CONCLUSIONS

In having demonstrated the evidence that irradiation of
any food commodity up to an overall average dose of 10 kGy
presents no toxicological hazard and introduces no special
nutritional or microbiological problem, the door is now open
for clearance of more irradiated foods in more countries.

The acceptance of the Recommended International General
Standard for Irradiated Foods (and the amendments presently
considered) by the Member States of the CAC can significantly
further the commercial introduction of food irradiation.
Acceptance of the Standard by many countries would also
facilitate the harmonization in the regulatory control of the
food irradiation process which is urgently needed to ensure
international trade of irradiated food.

Research and development programmes on food irradiation
are at present carried out in some 70 countries in the world.
Procedures for obtaining clearances and some forms of legislation
for the control of food irradiation have already been adopted
in 21 countries. These countries, however, have individual
legislative systems which differ in form and object of legislation, labelling provisions and in the organization of the
control of food irradiation. Both the absence of legislation
in many countries and the non-uniform system operative in
countries with legislation on food irradiation constitute
serious obstacles to the free flow of irradiated food in
international trade.

REFERENCES
1. WHO Technical Report Series, No 451, 1970. Wholesomeness
   of irradiated food with special reference to wheat, potatoes,
   and onions. Report of a Joint FAO/IAEA/WHO Expert Committee.
2. Ingram, M. Microbiology of foods pasteurised by ionising
   radiation. International Project in the Field of Food
   Irradiation, Karlsruhe, 1975. Technical Report Series
   IFIP-R 33.
3. WHO Technical Report Series, No 604, 1977. Wholesomeness
   of irradiated food. Report of a Joint FAO/IAEA/WHO Expert
   Committee.
4. Report of the eleventh session of the Codex Committee on
   Food Additives, The Hague, 31 May-6 June 1977. ALINORM
   78/12.

5. Legal Series No 11, IAEA, Vienna, 1979. International acceptance of irradiated food. Legal Aspects.
6. Codex Alimentarius Commission. Report of the twelfth session, Rome, 17-28 April 1978. ALINORM 78/41, Rome, 1978.
7. Report of the twelfth session of the Codex Committee on Food Additives, The Hague, 10-16 October 1978. ALINORM 79/12.
8. Report of the thirteenth session of the Codex Committee on Food Additives, The Hague, 11-17 September 1979. ALINORM 79/12-A.
9. Codex Alimentarius Commission. Report of the thirteenth session, Rome, 3-14 December 1979. ALINORM 79/38, Rome 1979.
10. WHO Technical Report Series, No 659, 1981. Wholesomeness of irradiated food. Report of a Joint FAO/IAEA/WHO Expert Committee.
11. Report of the fourteenth session of the Codex Committee on Food Additives, The Hague, 25 November-1 December 1980. ALINORM 81/12.
12. Revision of the Recommended International General Standard for Irradiated Foods and of the Recommended International Code of Practice for the Operation of Radiation Facilities Used for the Treatment of Foods. Report of a Consultation Group jointly organized by FAO/IAEA/WHO, Geneva, 1-3 July 1982. IAEA-TECDOC-258, Vienna, 1981.
13. Codex Alimentarius Commission. Report of the fourteenth session, Geneva, 29 June-10 July 1981. ALINORM 81/39, Rome 1981.
14. Takeguchi, Clyde A. Irradiated Foods - Criteria for deregulation. Paper presented at the first National Workshop on Radiation Processing of Food, University of Maryland, USA, 16 July 1981 (unpublished). (See also: Food Irradiation Newsletter 5 (No 3), December 1981.)
15. US Federal Register, Vol. 46, No 59, p.18992-18994 (27 March 1981). Policy for irradiated foods; Advanced notice of proposed procedures for the regulation of irradiated foods for human consumption.
16. Food Chemical News, 1981, 6 April, 4 May, 22 June, 6 and 20 July.

How does the consumer react to irradiated food?

Frederique Defesche

1. INTRODUCTION

In order to get an input from the consumer at the Symposium of 21 October, Gammaster asked Young & Rubicam-Koster b.v. to conduct qualitative research on the subject. The research had the objective of being an orientative study on the reactions and opinions of consumers in the matter of irradiated food and on the underlying attitudes. The method used was of a qualitative, small-scale nature using focus groups and depth interviews among Dutch housewives. The study must be seen as a preliminary piece of research and will be used as a pilot study. The results are in no way to be treated as factual data, they are "food for thought" and can only be indicative of Dutch reactions and attitudes.

As consumer or marketing research found its origin in psychological and communication theories. I would like to share two points of view with you:

1. Qualitative consumer research can only help in better understanding a problem. Its results are never absolute, never truly devoid of value judgements. Results can only become meaningful through the interpretation given to it. My paper is not always carefully worded. This is not because I am not aware of the relativity of what I am saying, but in order to make this speech more palatable to you.

2. The second point does sound absolute and in spite of what I just said on things being relative, this point is to be taken literally: "The consumer is right". He may from an objective and/or scientific point of view be wrong, yet the consumer's reality is reality as he perceives it selectively.

Food Irradiation Now – ISBN 90-247-2703-0
Copyright Gammaster and Martinus Nijhoff/Dr W. Junk Publishers

Many would argue - and especially those trained in science -
that such a reality is per definition distorted. Still I want
to maintain that the consumer is right. Right in his own
mind and feelings, and that is what we have to deal with.

2. WHAT DO WE TELL THE CONSUMER?

Alternative descriptions for the concept of food irradiation
were tested among consumers. Four alternative directions were
probed:
- irradiation of food
- transradiation of food (a term developed in order to see if
the new term would evoke the same or less negative reactions
than irradiation does)
- short-wave treatment of food
- electron treatment of food.

Spontaneous reactions to the term "irradiated food" showed:
- unfamiliarity
- dislike
- Angst in the psychological sense of an undefined fear
- fear of disease
- aversion because of unnaturalness.

The spontaneous reactions are primarily negative. The concept
"irradiated food" is unknown, people have no idea what it is,
yet their first reaction is one of rejection. However, it is
important to realize that these first associations do not go
in the direction of radioactivity. The idea of irradiated food
gives a first association with cancer, not with nuclear energy
or radioactivity. The term "irradiated food" without additional
information is incomprehensible, unclear and misleading to the
consumer, because it makes him draw wrong conclusions.

When consumers are confronted with another term for the
same concept the reactions are substantially more positive.
Consumers react to the term "transradiated food" spontaneously
by making a mental link to health checks, X-rays, sunbath.
Some think of microwave ovens or infra-red lamps. Transradiation is perceived to go right through without leaving
anything behind, whereas irradiation is felt to fall on top

of something leaving radiation in the product. "Transradiation"
is - of course - strange and unfamiliar, yet it is important
that the immediate association is not one of fear. The Dutch
word is "doorstralen" and far more friendly than its English
counterpart. "Doorstralen" means literally "shine through",
or "rays falling through".

Short-wave treatment or electron treatment of food is
incongruous in the consumer's mind. These words belong to
radios and machines, not to food. They simply cannot conceive
of short-wave treated food, except for astronauts. It seems
strange rather than scary. It sounds unnatural, far-fetched.

Again we may conclude that a single term describing the
irradiated food - whether we use the word irradiation or not -
is going to confuse if not mislead the consumer.

The next step in the pilot test was to confront consumers
with a short description of what irradiated/transradiated
food is. The descriptions did not explain the technical
process but rather the effect of it in highly simplified terms.
The descriptions used were for example:

- Transradiation/irradiation is one of the ways to render
bacteria and pathogens in our food harmless.

- Bacteria and pathogens exist in all the food we eat; they
can be rendered harmless with the help of a short-wave treatment.

3. WHAT DOES THE CONSUMER THINK WHEN THE BENEFITS AND ADVANTAGES
OF FOOD IRRADIATION ARE POINTED OUT TO HIM?

The question is: does this new information fit in with what
the consumer already knows and feels about food? And if it
fits in, how does it fit in, as this can be either positive
or negative in nature.

The consumer is immediately aware of the preserving aspects
of irradiation. The need to preserve food, however, is to
prevent spoiling during transport and for storage. Spoiled
food makes one ill. Consumers believe there are special bacteria
that cause the spoiling of food. They are not thinking of
pathogenic bacteria or viruses at all. Food only makes people

ill if it is kept too long or if one is allergic to it. That
food in an unspoilt state and looking and/or smelling fresh
can be the carrier of diseases is not realized.

The promise that food will remain fresh after transradiation and keeps because of it, is highly competitive as no
method of preservation exists that can claim the same. But it
will not be easy to communicate this convincingly to the
consumer. The message is confusing since the consumer believes
durability of food to be incompatible with freshness. To the
consumer freshness is closely linked to quality. Modern
consumers have lost the ability to judge the freshness or the
quality of the food they buy. In many cases it is packed and
cannot be seen, touched or smelled. Yet even if this could
be done the consumer has no way of telling what has been done
to food.

All processing and preservation methods are considered a
necessary evil. It is unnatural, therefore bad, but we cannot and do not want to go without the advantages. Distrust
of transradiated food must alone be seen in this perspective.
The consumer distrusts preservation methods, especially
chemical ones. Transradiation is a physical method with effects
on the food like pasteurization or sterilization. This is a
positive aspect of transradiation in the consumer's view.

The negative sides are (1) the association with irradiation
used to fight cancer and (2) fear of the use of a radioactive
source.

4. HOW IS FOOD TRANSRADIATED?

In the tests three basic ways were used to explain transradiation to the consumer:
1. the explicitly mentioning the use of a radioactive
Cobalt-60 source
2. not mentioning a radioactive source, but talking about an
electron or short-wave treatment or, about
3. mentioning particle accelerator ("van der Graaff versneller").
Neither the electron nor the particle accelerator story give
cause for fear. The radioactivity story does!

When considering the possibilities for introducing transradiated food on the consumer market, we ought to ask the
following questions:
- What should the consumer know?
- What does the consumer want to know?
Before we go further into this, we should stop and consider
what we mean by "to know". Knowledge and aspects of knowledge
are valued differently, depending for example on our training as
a psychologist or a physicist. Likewise the knowledge of a
consumer is not the same as the knowledge of a scientist. The
consumer's knowledge is perceived and selective. The fact that
he will usually only absorb what fits in comfortably with his
previous behaviour and attitude has its influence on if and
how he will accept transradiated food.

Right now, the consumer feels no need to trouble his head
about this new problem. And why should he! He has established
empirically and conclusively that he does not need food transradiation right now. He is still alive and kicking and does
in no way expect to drop dead any minute from the food he is
eating.

Does this imply that we should forget about food transradiation, because the consumer feels no need for it? No, not
necessarily! But we must fully realize and accept the fact
that the emotional significance of food is all powerful. The
knowledge aspects of consumer attitudes to food only constitute
a minor part of the total attitude. When selling food, one
hardly sells the means to satisfy a biological need; food is
likewise not consumed because of the technical and rational
explanation behind it.

Information regarding food should be objectively informative.
I will probably be offending quite a few people if I say that
this objectivity only sells well when it has at the same time
the added value of something on the affective level (such as
authority or confirmation). An additional problem in selling
food is that food generally shows a clear example of a "low
involvement" buying process, i.e., it is frequently purchased
at relative low prices and there is little time involved in

the buying decision itself. This means that consumers are not going to debate for hours if they are going to buy trans- radiated food. They are going to make up their minds pretty quickly, especially in front of the shelf in the shop.

All this adds up to careful and gradual introduction of transradiated food. A gradual step-by-step introduction of transradiated food makes it possible for the consumer to adapt gradually and for consumer information campaigns to start a learning process.

Step 1. Transradiation of packaging material before, during or after filling.

Step 2. Ingredients are transradiated instead of the finished product.

Step 3. The transradiation of those end-products that have at present such high contamination levels that they could be a danger to public health. In these cases transradiation should be recommended, if not made obligatory, by an official or government authority in order to give the consumer a guarantee from a higher and more neutral source than the manufacturer.

Step 4. Transradiation of end-products that are recognized by the consumer as health hazards (seafood in summer, raw meat, etc.).

Step 5. The transradiation of food products other than those mentioned above, for example ordinary daily food products but, only after a considerable period of adaptation and habituation. However, the products have to be accompanied by an additional claim or promise.

The transradiated food product will have to compete on the shelf with non-transradiated products. The consumer is going to decide in seconds. Not only must the description not cause fear, it must promise something that appeals to the consumer emotionally and immediately. Emotionally in a positive way. Phrases like "irradiated food, germ-free" will not do the job. They only make the consumer think of disease, both because of the word "irradiation" and because of the word "germ". Neither do statements such as "irradiation does not cause the product to be radioactive" or "irradiation is entirely without danger" tell what it intends to tell. In fact they do more harm than good. What do these statements tell the consumer? It is talking about radioactivity and danger and specifying this with the additions "not" and "without".

A simple exercise will help to prove this point. For example, try not to think of a house. Concentrate! Do not think of a house! What have you been thinking of, doing this exercise? What mental picture did you get? Right, not a non-house. No such thing exists. You were mentally seeing a house (either visually or spelled out in letters).

Hopefully this has made it clearer why it is necessary to make the emotional promise a positive one, especially on a food label where the confrontation is short.

Several claims/promises were probed during the test:
- irradiated food is safe and fresh food
- transradiated food is fresh and safe food
- transradiated food remains fresh longer and is safe food
- electronically purified food
- fresh and safe thanks to short-wave purification

Claims or emotional benefits such as fresh, safe, pure and purified create a positive result on the affective level of consumer attitudes. Emotional benefits such as "keeps longer" and "more wholesome" or "healthy" are less direct. "Keeps longer" in itself is not a benefit; the consumer can mentally compute this to "easy, less frequent shopping", etc., but the benefit does not cater to an immediate emotional need. The same is true for the claim "wholesome". Wholesome or healthy is only a direct emotional promise to the sick. To the average healthy person a promise of "feeling good or looking good" is more direct.

Next to a positive claim an additional mention on the packaging that this method has been approved by the Ministry of Agriculture or by the WHO would create (more) confidence and promote consumer acceptance. The guarantee or recommendation does not only work on a rational level, but is in addition an emotional promise to the consumer. In this, the promise is an extra quality expectation created by a guarantee from such reliable sources.

Overall consumer attitudes to transradiated or irradiated food fall into two categories, positive and negative ones. The negative attitude shows the following aspects:

1. The consumer wants to avoid being confused and alarmed and rejects the confrontation with unpleasant problems.

2. Former experience (personal or indirect) with the link between disease/death and radiation by means of a radioactive source.

3. Fear of radioactive residues in the food and of becoming radioactive oneself through an accumulative use. The term "transradiation" evokes less fear since "trans" means: going right through and not leaving residues.

4. Devitalization of food. The fear is that by slowing down the ability to germinate or ripen, the food is "incapacitated", is becoming a "food invalid" left without any "goodies" (brown bread may look brown but has the nutritional value of "dead" white bread).

5. Denaturalization of food. The fear that the food has been killed, whereas food is perceived to be alive. In the consumer's mind vitamins, minerals, protein, etc. are all alive. How could they do any good if they were not?

6. Fear that control at the point of irradiation is insufficient.

7. Fear that methods of disposal of waste materials are unsafe.

The positive reactions are the following:

1. Consumers are interested in fresh and safe food and feel the need for it.

2. Interest is shown in rendering safe of those foods that are known to be dangerous.

3. In spite of the incomprehensible contrast, an interest in
- durable and fresh
- safe and raw.

4. Purified or pure food without (chemical) additives has a great attraction, mainly on a psychological level.

5. Confidence in the recommendations and guarantees from official and government agencies.

5. CONCLUSION

Our conclusion is that there are ways that could lead to consumer acceptance, provided:
- steps are taken gradually
- negative and misleading descriptions are avoided
- that on the way to acceptance emotional benefits and advantages are offered to the consumer.

There is as yet no real proof that the consumer does not accept transradiated food. His fear is not focused on the transradiated food, but on cancer and radioactivity. Besides, the consumer is neither aware of the health hazards of our food nor of the enormous losses suffered in storage and transportation. And how could he be aware of it? Virtually anything that has to do with food production, from processing, preserving, packaging to transportation, storage and quality control is totally hidden from his view.

We live in a society where all these provisions are handed to the consumer as a matter of course. It is, therefore, unrealistic to wait until that same well-protected and cared-for consumer takes the initiative in favour of transradiated food by making it clear to you that he feels the need for it.

Panel discussion

After the individual presentations the authors formed a panel which answered written and oral questions from the audience. The questions and answers are grouped in subjects omitting the names of addressors and advisers. The subjects are:
- microbiology;
- technology;
- legal aspects;
- consumers' aspects.

1. MICROBIOLOGY

Q. What causes the destruction of bacteria, moulds or yeasts by radiation, and how rapidly does this occur?

A. Microorganisms are killed at the cellular level: the DNA strain is disrupted almost immediately.

Q. How do you measure the required dose?

A. One measures the effect of increasing doses of irradiation on pure cultures of microorganisms or of emulsions of organisms in foods. From such dose response curves D_{10} values are calculated.

Q. The dairy industry succeeded in overcoming their microbiological problems without radiation. Do you see a future here for this technology?

A. There are three answers to this rather complicated question:
1. Because irradiation of foods with relatively high protein and fat contents at ambient temperatures may cause off-flavours,

the dairy industry has been rather reluctant to apply this
technique.

2. Cartons for sterilized milk are in some instances irradiation-
disinfected before aseptic filling.

3. In some dairy products such as cheese a certain population
of microorganisms is required.

Q. What possibilities do you see for the irradiation of
catering products and convenience foods?

A. If storage life is only a function of microbial contamination there is certainly a future. If, however, enzymatic
processes play a dominant role irradiation alone can do little.

Q. What influence does irradiation have on certain food
contents such as vitamins?

A. The most irradiation-sensitive vitamin is ascorbic acid
(vitamin C). The degradation at low irradiation doses is
comparable to that during cooling; over 10 kGy it is comparable
to cooking. If irradiated food is the main substance of the
diet, the Joint FAO/IAEA/WHO Expert Committee on the Wholesomeness
of Irradiated Food (Technical Report Series 659, 1981) recommends
an investigation on the influence of irradiation on that food
item specifically and to the complete diet generally.

Q. Is it considered adequate to eliminate <u>Salmonella</u>
populations from fish products with a dose of 3 kGy?

A. Experimentally it was found that the D_{10} of most <u>Salmonella</u>
strains equals 0.6 kGy. At 3 kGy the reduction will be 5 decay
(10^5), at 4 kGy even 7 decay. This holds to a slightly lesser
degree when the product is deep-frozen (-12 $^\circ$C or lower).

Q. If the application of irradiation becomes widespread,
what will be the chances for radioresistant microorganisms
to emerge? Compare this situation with what has happened within
hospitals with regard to antibiotics.

A. In laboratory experiments with doses in kGy range it was
possible to isolate some radioresistant strains. But fortunately
they were unable to compete with sensitive strains in the
ambient biosphere. So this risk seems negligible. Moreover,
most food products are pre-packed before treatment. This
implies that irradiated microorganisms are eliminated from

the source, so that recirculation is impossible, unlike the situation in hospitals.

Q. Is it feasible to identify a food as having been irradiated? According to Kampelmacher and Tamminga irradiated food has the microbial status of pasteurized food and the enzymatic status of fresh food.

A. This approval is fully valid. However, this monitoring technique requires at least 24 hours. Hence for instance when fresh such as perishable products have to be cleared for importation, this time delay may render monitoring impracticable.

Q. Are viruses affected by irradition?

A. Yes they are but the majority are extremely radiation resistant.

2. TECHNOLOGY

Q. In an earlier question the problem was raised that high doses cause deterioration of proteins (degradation) and fats (oxidation). The food may become organoleptically unacceptable. This is caused by the radicals formed. Does this organoleptic loss of quality coincide with a toxicological problem?

A. The toxicological research applies the dose-response relation. At the upper end there is a certain exaggeration, at the lower end there is a no-effect level. The ratio between the two is the safety factor and explains which should be the dose limits for consumption. This general approach does, however, not hold for irradiation because if too high doses are given, the food changes in composition just as by over-cooking or overheating. Hence, the percentage of the irradiated food as part of the complete diet was increased. This has been carried out in feeding tests with pigs. Even so the noxiousness of irradiated food could not be demonstrated successfully: it turned out to be toxicologically safe, such as wholesome.

Q. Are all packaging materials suitable for irradiation?

A. All materials are changed by irradiation. The levels of alteration, however, are acceptable. For convenience purposes the Dutch authorities apply the same list of accepted packaging materials as cleared by the Food and Drug Administration in 1967.

Q. How long is the life of a Co-60 source? What eventually happens to it? Does the radiation only consist gamma rays?

A. The activity of a source is determined by the half-life of the Co-60. This is about 5.2 years or 1% per month. The Co-60 is present in stainless steel pencils. When such a pencil gets too low in activity it is extracted from the source and shipped to the manufacturers, who have a choice of two possibilities: either to reactivate the Co-60 or to use it as such in a low-activity research source. Because of the encapsulation of the Co-60 in these pencils most of the accompanying β-rays are immediately absorbed by the stainless steel. Those with an energy sufficiently high to penetrate through the capsule cause the well-known blue light (Tsjerenkow effect), which may be seen when the source is under water.

3. LEGAL ASPECTS

Q. What determines the viability of a product for irradiation? A bacteriologically unacceptable product, for example with a total count of over 10^6 - 10^8 microbes per gram, could be made acceptable by irradiation.

A. This example is almost hypothetical. The bacterial contamination is self-limiting. The required reduction dose is so high in these instances that organoleptical problems may be expected. A bad product can never become a good one by irradiation.

Q. In the Netherlands there are more than a dozen clearances for irradiated commodities. Where can one find these?

A. They are all published in the "Staatscourant", and also in other publications. The Ministry of Health is the best information source for anyone interested.

Q. If irradiated products form only part of the ingredients of a food, must this be declared on the labelling?

A. In this case irradiated food is treated in the same way as chemically preserved, gased or frozen products: none of them need be declared. If, however, the enteric contents are treated, the declaration is carried out by the irradiation symbol. This symbol is the property of the "Proefbedrijf Voedselbestraling" at Wageningen but the authorities are licenced to apply it or have it applied by other firms. At the moment negotiations are going on to get the symbol internationally accepted as the declaration label.

Q. How do other countries around the Netherlands, for example Germany, accept irradiated foods?

A. The German Federal Republic (BRD) is a long way behind the Netherlands because the German food industry has not been very active. It eliminated the question of clearances. The one exception being the permission granted to the "Förderkreis für Lebensmittelbestrahlung", Bremerhafen, which received permission to treat fish products with an on-board X-ray irradiation source.

Press releases declaring that there is a vivid export of irradiated chicken to Germany are completely unfounded since neither of the two irradiation facilities in the Netherlands have received any chicken for treatment during the past years. Only a few samples were treated for test purposes.

Q. Why are clearances with all their legal obligations necessary at all since international organizations consider irradiated food as safe?

A. As always the law lags behind. During the past two decades a toxicological and microbiological control was considered necessary. Those positive and encouraging declarations towards the irradiated product are barely two years old. It will take some time before it is generally and unconditionally accepted. Until then the system of clearances will be required.

4. CONSUMERS' ASPECTS

Q. Without sufficient and independent information the acceptance of irradiation of food is uncertain. Who will supply this information?

A. This is an open question for the time being. The public

expects this type of information from the authorities, the
Voorlichtingsbureau voor de Voeding (Netherlands Bureau for
Food and Nutrition Education), perhaps from TNO (Netherlands
organization for applied scientific research). Because irradiation is such a fundamentally new method, the consumer would
like a guarantee from a neutral source. The information cannot
come from labelling alone. Labelling can only provide limited
information and extreme caution is necessary here as limited
information can be interpreted wrongly. As the subject is
complicated the information should be phased in order to facilitate consumer understanding. Again this is not a problem for
one single country, but should be solved by more nations.

Q. Can the consumer understand the technical information,
as it is extremely complicated in this case?

A. We must prepare ourselves to answer consumer questions
on consumer problems, meaning that the problem-areas must
be translated into areas that the consumer sees as a problem.
As such consumer acceptance does not depend solely on technical
information. The consumer must gradually become familiar with
the benefits of preservation by means of irradiation. Not only
such vague benefits as public health an the world food supply.
The more direct the benefit is to the individual consumer, the
sooner will there be acceptance.

Q. How can consumer acceptance be stimulated?

A. By phasing both the type of consumer foods that will be
irradiated an the information about it. First step is packaging
and ingredients that pose a health problem. Then endproducts
that are a health hazard according to the authorities. Maybe
a guarantee from them should accompany such a product in the
initial stages. Endproducts that spoil quickly and noticeably
to the consumer can come next. Also, if irradiation can result
in a cheaper product for the consumer - because of less waste,
cheaper packaging, less time, etc. This would be an immediate
consumer benefit that will not fail to translate itself into
acceptance.

Q. The word "irradiation" is such a loaded word. The dutch
word "doorstralen" has been tested and seems to have less

negative connotations. This word may not be translatable into other languages.

A. True, the word "irradiation" is highly charged. As long as an average consumer has not had the opportunity to learn and understand that the irradiation of food is not going to irradiate him, food irradiation must be treated as a new concept A new concept deserves a new word. As long as irradiation means ill people and nuclear plants, the word in relation to food is far from being informative. In fact it is downright misleading. We must look for new ways to describe and explain this new concept better and more truthfully. The dutch word "doorstralen" was only an example, it could also be achieved by other words, sentences, visual symbols, colours, etc.

Wholesomeness of irradiated food

Report of a Joint FAO/IAEA/WHO Expert Committee

This report contains the collective views of an International group of experts and does not necessarily represent the decisions or the stated policy of the Food and Agriculture Organization of the United Nations, the international Atomic Energy Agency of the World Health Organization.

World Health Organization
Technical Report Series
659

CONTENTS   Page

118

JOINT FAO/IAEA/WHO EXPERT COMMITTEE ON THE WHOLESOMENESS
OF IRRADIATED FOOD

Geneva, 27 October - 3 November 1980

Members:

Dr H. Blumenthal, Director, Division of Toxicology, Bureau
of Health and Human Services, Public Health Service, Food
and Drug Administration, Washington, DC, USA.

Dr B. Briski, Chief, Division for Laboratory Food and Food
Additives Analyses, Department of Nutrition, Institute of
Public Healt of Croatia, Zagreb, Yugoslavia.

Professor D.O. Cliver, Food Research Institute and Department
of Food Microbiology and Toxicology, University of Wisconsin,
Madison, WI, USA.

Professor J.F. Diehl, Director, Institute of Biochemistry,
Federal Research Centre for Nutrition, Karlsruhe, Federal
Republic of Germany.

Dr J.C. Gould, Director, Central Microbiological Laboratories,
Western General Hospital, Edinburgh, Scotland (rapporteur).

Dr M. Ishidate, Jr, Director, Division of Mutagenesis,
Biological Safety Research Centre, National Institute of
Hygienic Science, Tokyo, Japan.

Dr M. van Logten, Head, Laboratory for General Toxicology,
National Public Health Institute, Bilthoven, Netherlands.

Professor M. Mercier, Head, Laboratory of Biotoxicology,
Faculty of Medicine and School of Pharmacy, Catholic Uni-
vers-ty of Louvain, Brussels, Belgium.

Dr A.O. Olorunda, Department of Food Technology, Faculty of
Technology, University of Ibadan, Ibadan, Nigeria.

Professor M.J. Rand, Chairman, Department of Pharmacology,
University of Melbourne, Parkville, Victoria, Australia
(chairman).

Mr. L. Saint-Lèbe, Radiation Conservation of Foodstuffs
Group, Radiation Agronomy Service, Cadarache Centre for
Nuclear Studies, St Paul-lès-Durance, France.

Dr P.G. Tulpule, Director, National Institute of Nutrition, Indian Council of Medical Research, Hyderabad, Andhra Pradesh, India (vice-chairman).
Dr K. Vas, Director, Central Food Research Institute, Budapest, Hungary.

Observers (invited by FAO/IAEA):
Dr J. Farkas, Project Director, International Facility for Food Irradiation Technology (IFFIT), Wageningen, Netherlands.
Mr. A Feberwee, Chairman of Codex Committee on Food Additives, Ministry of Agriculture and Fisheries, The Hague, Netherlands.
Mr. W.T. Potter, Project Secretary, International Food Irradiation Project, Nuclear Energy Agency, Organization for Economic Cooperation and Development, Paris, France.

Secretariat:
Mr. W.R. Bradford, Principal Scientific Officer, Food Science Division, Atomic Energy Branch, Ministry of Agriculture, Fisheries and Food, London, England (FAO/IAEA Temporary Adviser).
Dr A. Brynjolfsson, Head, Radiation Preservation and Food Division, Food Engineering Laboratory, US Army Natick Research and Development Laboratory, Natick, MA, USA (FAO/IAEA Temporary Adviser).
Dr P. Elias, Project Director, International Project in the Field of Food Irradiation, Federal Research Centre for Nutrition, Karlsruhe, Federal Republic of Germany (FAO/IAEA Temporary Adviser).
Dr K.O. Herz, Food Standards and Food Science Service, Food Policy and Nutrition Division, FAO, Rome, Italy (FAO Joint Secretary).
Dr F.K. Käferstein, Responsible Officer for Food Safety, Unit of Environmental Hazards and Food Protection, Division of Environmental Health, WHO, Geneva, Switzerland.
Dr W. Keller, Nutrition Unit, Division of Family Health, WHO, Geneva, Switzerland.

Mr. J.G. van Kooij, Head, Food Preservation Section, Joint
FAO/IAEA Division of Isotope and Radiation Applications of
Atomic Energy for Food and Agricultural Development, IAEA,
Vienna, Austria (IAEA Joint Secretary).

Dr A Koulikovskii, Veterinary Public Health Unit, Division
of Communicable Diseases WHO, Geneva, Switzerland.

Dr L.G. Ladomery, FAO/WHO Food Standards Programme, FAO,
Rome, Italy.

Dr F.C. Lu, Consulting Toxicologist, Miami, FL, USA (WHO
Temporary Adviser).

Dr N.T. Racoveanu, Chief, Radiation Medicine, Division of
Noncommunicable Diseases, WHO, Geneva, Switzerland.

Professor H. Roushdy, Director, National Centre for
Radiation Research and Technology, Atomic Energy Authority,
Cairo, Egypt (WHO Temporary Adviser).

Dr K. Sundaram, Director, Division of Life Sciences, IAEA,
Vienna, Austria.

Dr G. Vettorazzi, Food Toxicologist, International Programme
on Chemical Safety, Division of Environmental Health, WHO,
Geneva, Switzerland (WHO Joint Secretary).

Dr V. Volodin, Radiation Medicine, Division of Noncommunicable
Diseases, WHO, Geneva, Switzerland.

WHOLESOMENESS OF IRRADIATED FOOD

Report of a Joint FAO/IAEA/WHO Expert Committee

A Joint FAO/IAEA/WHO Expert Committee on the Wholesomeness
of Irradiated Food met in Geneva from 27 October to 3
November 1980. The meeting was opened by Dr T. Fülop,
Director of the Division of Health Manpower Development, on
behalf of the Directors-General of the Food and Agriculture
Organization of the United Nations, the International Atomic
Energy Agency, and the World Health Organization. He mentioned
that, as a result of recommendations from previous Joint
Expert Committees and the conclusions of other technical or
legal expert consultations organized by these agencies, the
FAO/WHO Codex Alimentarius Commission had adopted a general
standard for irradiated foods as well as a code of practice
relating to food irradiation facilities. Once the recommended
general standard is accepted by Governments, foods evaluated
by the Expert Committees would be permitted to be irradiated.
These would include chickens, papaya, potatoes, strawberries,
wheat and ground wheat products, cod and redfish, onions,
rice, mangoes, dates, cocoa beans, spices, and pulses. A
number of these products is of special interest to developing countries.

1. INTRODUCTION

The world's food requirements continue to grow, but in an
environment of scarce resources and of limitations on methods
of food production. In addition, the problems of food storage
and processing make it necessary to search for effective
alternative methods of food preservation, particularly where
existing methods are costly because of the energy requirements
and may be difficult to provide in some areas.
Accordingly, it is reasonable to consider the use of ionizing
radiation for food storage and preservation as one alternative,
provided that it does not adversely affect the wholesomeness

of food.

The need to consider the wholesomeness of food processed by irradiation was emphasized at an international level at a meeting sponsored by FAO, IAEA and WHO in Brussels in 1961 (1). The studies required to ascertain the wholesomeness of irradiated food were discussed by a Joint FAO/IAEA/WHO Expert Committee on Irradiated Food in Rome in 1964 (2). Taking as a premise that the irradiation of food resulted in the production of radiolytic products in the food, the Committee adopted the view that these products represented additions to the food. It therefore concluded that the establishment of the safety of irradiated foods should follow procedures similar to those generally used for evaluating the safety of food additives and should be pursued on a food-by-food basis.

A subsequent Joint Expert Committee, which met in 1969 (3), had available for consideration the result of a number of toxicological studies carried out on three specific foods on the basis of the recommended procedures. It reviewed the comparative data on several varieties within a major crop, and accepted extrapolation of data from a major variety to all varieties of that crop. The Committee recommended temporary acceptance of irradiated wheat and potatoes as wholesome, and specified further studies on onions. The next Joint Expert Committee, convened in 1976 (4), reviewed a large number of animal studies on various irradiated foods. Unconditional or provisional acceptances were recommended for most of them. The Committee also reviewed the results of radiation chemistry studies on the major components of food; it noted that many of the radiolytic products identified were present in food treated by heat and other processes and considered that the health hazard from the concentrations found was probably negligible. It therefore encouraged further studies on the chemical changes in food components associated with irradiation.

A large number of data on irradiated foods and food components have since been generated. The present Committee was convened to evaluate the wholesomeness of the irradiated foods for which data were available. It was also asked to review the acceptability of irradiated food in general, in the light of all the toxicological data and the data from radiation chemistry studies, and to make suggestions for further studies where desirable.

2. GENERAL CONSIDERATIONS

2.1. Principles
The principles and guidelines set out in the reports of the 1964, 1969, and 1976 Joint FAO/IAEA/WHO Expert Committees formed the basis for the present Committee's approach to its consideration of the wholesomeness of irradiated food.

2.2. Reasons for the use of food irradiation
The Committee was aware that irradiation of food may be used to achieve a variety of desirable objectives including the following, which are classified according to the average radiation dose required to achieve the objectives in question:

Low-dose applications (up to about 1 kGy)
Inhibition of sprouting
Insect disinfestation
Delay of ripening

Medium-dose applications (about 1-10 kGy)
Reduction of microbial load
Reduction in the number of non-sporing pathogenic micro-organisms
Improvement in technological properties of food

High-dose applications (about 10-50 kGy)
Sterilization for commercial purposes
Elimination of viruses

The sections that follow (3-7) summarize the evidence which enabled the Committee to assess the effect of the ir-

radiation process on the wholesomeness of food and to arrive
at conclusions on the acceptability of irradiated foods.

3. TECHNICAL ASPECTS

3.1. Radiation sources

The Committee stressed the importance of using appropriate
radiation sources. From the point of view of safety, the
energy level of the radiation applied to food is the most
important characteristic that has to be regulated in order
to prevent the possible formation of induced radioactivity
in the irradiated material. In practice, this is only of
importance when considering machine sources, since the most
commonly used isotopic sources (^{60}Co and ^{137}Cs) emit radiation
of a maximum energy ($\leq$ 1.33 MeV) which is lower than that
causing induced radioactivity. The Committee examined a
recent unpublished report (5) showing that, with machine
sources, induced activity is negligible and very short-lived
below an energy level as high as 16 MeV. In this respect the
Committee reconsidered and endorsed a statement (in the
report of a Joint FAO/IAEA Advisory Group on International
Acceptance of Irradiated Foods (6)) that the radiation permitted for food irradiation should have a maximum energy
level of (a) 10 MeV for electrons and (b) 5 MeV for gamma
rays and X-rays. On the basis of that statement and the
report of the Expert Committee that met in 1964, which
indicated X-rays as a suitable type of radiation, the present
Committee decided to recommend the inclusion of X-ray-
sources in the list of acceptable radiation sources.

3.2. Absorbed dose

The present Committee reiterated the view of the Expert
Committee of 1976 (4) that, as a matter of principle, the
applied dose of ionizing radiation should not be higher or
lower than is needed to achieve the desired effect. Finding
and applying the appropriate dose level is the key to the
technologically and economically proper application of the

irradiation process to food.

It was stressed that the application of the correct dose would be taken care of, wherever there was good irradiation practice. It was recognized that advice on the doses necessary for the treatment of specific food items and the procedures involved would assist those concerned. Such advice could be included in a code of technological practice.

The Committee noted that no new method for the determination of absorbed dose in the food itself, or indeed for the identification of irradiated food, had become available since 1976. It therefore upheld the view of the Expert Committee that met in 1976 (4) that effective dose control can only be exercised in the irradiation plant. The operation of irradiation facilities should be subject to supervision by the appropriate national authorities in order to ensure that proper dose control is exercised. In this respect it was noted that assistance in the calibration of dose control is offered by the IAEA through its programme on High- and Low-Dose standardization and inter-comparison for industrial radiation processing.

As regards setting an overall average dose[1] for the process of irradiation, it was considered that, contrary to the opinion expressed by the Expert Committee that met in 1976 (4), it is practical (for reasons such as the technical design of the irradiation facility) to stipulate an average value rather than to require that no part of the food shall receive less than a minimum, or more than a maximum dose. Taking into account the ratio of maximum to minimum dose

-----

1. The overall average dose is the arithmetic mean value of all dosemeters readings in a given irradiation run. To determine this mean value, an adequate number of dosemeters must be randomly distributed in the food as it is exposed to the radiation. The number of dosemeters is considered adequate if it permits estimation of the dose distribution in each portion of the food material of different density and if the measurements are representative for all dose and density fluctuations during a usual run.

absorbed by the product (i.e., the "dose uniformity ratio")
in pilot and currently used commercial facilities, the overall average dose may result in a small friction of the food
receiving a maximum absorbed dose up to 50% higher.

3.3. Processing conditions for irradiation

It is expected that, with wider application of food
irradiation, processing conditions will be designed to meet
specific technological requirements. Plant design should
attempt to minimize the dose uniformity ratio to ensure
appropriate dose rates and, where necessary, to permit
temperature control during irradiation (e.g., for the
treatment of frozen foods) and also control of the atmosphere.
It is also necessary to minimize mechanical damage to the
product during transportation, irradiation, and storage, as
well as to ensure the maximum efficiency in the use of the
irradiator. Where the food to be irradiated is subject to
special standards for hygiene or temperature control, the
facility must permit compliance with these standards.

3.4. Packaging of irradiated food

The packaging method and the packaging material used must
be safe and appropriate to the food to be irradiated.
Irradiation must not adversely affect the functional properties of the material chosen, nor must it render the material
unsafe as determined by appropriate test methods of the kind
applied to the unirradiated material.

3.5. Repeated irradiation

While adhering to the view that irradiation of food should
normally be carried out once only in each case, the Committee
agreed that in certain circumstances repeated irradiation
might be justified. This is a departure from the statement
in the report of the Expert Committee that met in 1976 that
any repetition of irradiation is to be avoided. In deciding
upon this change, the present Committee took account of the
following findings: (a) the concentration of radiolytic

products is a linear function of dose; (b) there is a considerable and rapid reduction in the concentration of some
of these radiolytic products following irradiation; and (c)
an overall average dose based on toxicological and other
considerations could now be established (see section 10).
Consequently, a repetition of irradiation within this overall average dose would not be harmful, provided that no
significant impairment of nutritional or technological
properties occurred. The Committee agreed that, at the present
stage of knowledge, the acceptability of repeated irradiation
should be limited to the case of food commodities of low
moisture content, in which reinfestation by insects could not
be effectively prevented under practical conditions of
storage and transport.

Two other types of repetition of the irradiation process
were also considered acceptable: (a) when the food to be
irradiated is a processed form of food that has already
undergone low-dose treatment (for example, dried onion
prepared from onions treated to inhibit sprouting); (b) when
it includes irradiated minor ingredients (for example, meat
products or dehydrated soup containing irradiated spices).
In both cases it was considered that the additional amounts
of radiolytic compounds formed in the final products would
be insignificant.

By analogy with tyndallization, fractionated irradiation
(i.e., when the full dose has to be applied in two or more
instalments) should not be considered as repeated irradiation.

3.6. Technological efficacy

The Committee stressed that, like other food processing
techniques, food irradiation is justified only if it serves
a useful purpose. Results of studies on the efficacy of the
irradiation of the food items specifically examined by the
present Committee clearly showed that the applications in
question are technologically justified and effective.

3.7. Requirements of quality assurance and labelling

The use of sound raw materials and proper handling and
processing techniques, as well as strict maintenance of the
wholesomeness and other desirable qualities of foods are a
necessity when irradiation or any other form of processing
is applied. Furthermore, users and consumers are entitled
to expect that the quality and safety of food is not adversely changed either by irradiation or by other currently
accepted forms of treatment.

The Committee understood that irradiated foods would be
subject to regulations covering foods generally, and to any
specific food standards relating to individual foods. It was
therefore not thought necessary on scientific grounds to
envisage special requirements for the quality, wholesomeness,
and labelling of irradiated foods.

4. ASPECTS OF RADIATION CHEMISTRY

4.1. Chemical analysis and wholesomeness evaluation

Treatment of foods with electrons (of energies up to
10 MeV) or gamma-rays and X-rays (of energies up to 5 MeV)
does not produce radioactivity in the foods so treated. The
need for toxicological evaluation of irradiated foodstuffs
stems from the fact that the application of radiation energy
results in chemical changes. The nature of the radiation-
induced compounds depends primarily on the chemical composition of the food. The concentration of radiation-induced
compounds generally increases with increasing radiation dose,
but can be modified by factors during irradiation such as
temperature, presence or absence of air, and the water content
of the sample. The energy taken up by the irradiated food is
much less than that taken up by heated foods. It is therefore
not surprising that chemical changes caused by irradiation
are quantatively much smaller than those caused by heating.
For instance, an absorbed dose of 10 kGy (1Mrad) corresponds
to a temperature rise of only 2.4° C in a food having the
heat capacity of water (4.184 $J/^{\circ}C$; 1 $cal_{th}/^{\circ}C$). This is

about 3% of the energy needed for raising the temperature of
water from about 20° C to 100° C.

The Expert Committee that met in 1976 concluded that the
radiolytic products detected in the wide range of foods and
individual food constituents that had been studied did not
appear to pose any toxicological hazards in the concentrations
at which they were detected. That Committee also accepted
that, for doses below 10 kGy (1 Mrad), data may be extrapolated from one member of a food class to related members
(p. 10 in that Committee's report (4)) and, furthermore, that
if certain studies in radiation chemistry and toxicology were
continued, a purely chemical approach to the wholesomeness
evaluation of irradiated food may prove to be possible (p. 11
in the report (4)).

4.2. Recent studies

The above proposals stimulated a great deal of chemical
research on irradiated foods and on model systems, which has
confirmed the earlier assumptions and enabled more radiolytic
products to be identified and quantatively determined. Thus,
the mechanisms of radiation chemical reactions in carbohydrates,
lipids and proteins are now known in greater detail.

A study of the radiolytic products in beef, pork, ham and
chicken has shown that formation of volatile hydrocarbons
depends on the fat content of the meat, regardless of origin.
The electron spin resonance spectra from the four types of
meat irradiated at -40 $^\circ$C were identical, indicating the
production of common free radical intermediates (I.A. Taub &
C. Merritt, unpublished observations).

Another study showed radiolytic products from various
starches (derived from maize, amylomaize, waxy maize, wheat,
manioc, potatoes, rice, and beans) to be qualitatively
identical. Small quantitative differences were related to
known properties of these starches, such as the ratio of
amylose to amylopectin. These results were confirmed by
electron spin resonance which showed that the nature of the
radical intermediates is the same in all the irradiated

starches (J. Raffi & L. Saint-Lèbe, unpublished observations).

A study of radiation-induced changes in a fruit model has shown that the extent to which these changes take place is in accord with well established kinetic laws. These changes may be calculated using digital computer methods to solve the differential equations which describe the reaction probabilities. Chemical analysis confirmed the prediction that the radiolytic products present in greatest yield in the irradiated fruit were derived from the major constituents of the fruit, i.e., from sugars. Yields of products derived from minor constituents such as protein, malic acid, phenolics, and nicotinamide were much lower (R.A. Basson and co-workers, unpublished observations).

The products of radiolysis in beef (irradiated with an average dose of 56 kGy (5.6 Mrad) at -30 $^{\circ}$C) have been studied in detail. Over 100 volatile compounds have been identified at concentrations varying from 1 to 700 µg/kg, with a total yield of 9 mg/kg. Most of the compounds are known to occur also in unirradiated foods. The Committee noted that this subject had been reviewed recently (7, 8) and agreed that there were no grounds for suspecting these products of being a hazard to the consumer.

4.3. Conclusions

Since similar radiolytic reactions occur with the same constituents of different foods (protein, fat, carbohydrates, water, etc.), common radiolytic products are formed in roughly predictable yields when these foods are irradiated. Although only approximate predictions of product yields are possible at present, these are sufficiently accurate to enable estimates to be made of the upper limits of yields. Thus there is now considerable additional evidence to support the view that information obtained from toxicity tests on one irradiated food can be extrapolated to other foods of similar chemical composition, or to other processing conditions for the same food.

5. NUTRITIONAL ASPECTS

None of the evidence published since 1976 necessitates a
change in the advice on the nutritional aspects of irradiated
food given by the Joint Expert Committee that met in that
year (4). The salient points are as follows:

Evidence from most studies suggests that in the low-dose
range (up to 1 kGy) used for the irradiation of food, nutrient
losses are insignificant. In the medium-dose range (1-10 kGy),
losses of some vitamins may occur, if air is not excluded
during irradiation and storage. In the high-dose range (10-
50 kGy), the technology used to avoid effects on organoleptic
quality (i.e., irradiation at temperatures below freezing
and in the absence of air) also partially protects nutrients,
so that losses may actually be lower than in the medium-dose
range if such precautions have not been taken.

Conflicting results have been reported concerning the
effect of radiation on vitamin C levels in foods. Some
authors have determined only ascorbic acid to dehydro-
ascorbic acid, which is also biologically active. In future
studies, both ascorbic and dehydro-ascorbic acid should
therefore be determined.

The extent of losses of nutrients due to the irradiation
of foods depends on many factors, such as the composition of
the food, the radiation dose, the temperature, and the
presence or absence of air during irradiation and storage.

Whether or not the loss of a nutrient in an irradiated
food is of importance depends on circumstances, such as the
contribution that this food makes to the total diet. For
instance, a partial loss of thiamine in fish would be of
concern if that was the key source of thiamine to a particular
population. Other relevant factors include the nutritional
status and requirements of the population for which that
food is intended. Some other areas of uncertainty (i.e.,
folic acid losses) require further investigation.

In 1976 the Joint Expert Committee suggested that the
reduction of nutritional value produced by irradiation alone

should be compared with that produced by other processes (4). A considerable body of evidence is now available in this regard and the results give no cause for particular concern.

6. MICROBIOLOGICAL ASPECTS

The microbiological safety achieved by the food irradiation process is fully comparable with that of other currently accepted food treatments. No findings have been published during the past four years which would necessitate a reconsideration of the views expressed by the Joint Expert Committee in 1976 (4) regarding the microbiological implications of irradiation of food. The results of theoretical and practical work carried out since 1976 have not revealed any new microbiological problems besides those already reviewed.

The results of both field and "inoculated pack" studies have shown that the microbiological safety evaluation of a specific irradiated food can be based only on studies that have specifically been designed to reflect all the circumstances encountered in commercial irradiation. Furthermore, it is important that the hygienic aspects of each individual commodity should be examined separately and that the post-irradiation storage conditions should be carefully and adequately designed to control microbial growth.

6.1. Variations in radiation resistance

The natural radiation resistance of microorganisms and the consequences of their possible survival after irradiation have been investigated with regard to some highly radiation-resistant microorganisms. No new health hazards arising from these organisms have been identified.

Additional experience has also been gained in the application of potentially useful and technologically acceptable combined treatments. For example, it has been demonstrated that the use of irradiation, in conjunction with heat and/or salt treatment, achieves a more efficient reduction in the number of organisms, especially the highly radiation-resistant organisms.

6.2. Radiation-induced genetic variations

Since 1976 there have been no reports to justify the concern, expressed before that time, about the development of irradiation-induced mutations under good operating conditions. As already stated in 1976 (4), the risk of inducing greater radiation resistance has only been shown under laboratory conditions.

Changes of taxonomically relevant characteristics, due to mutation, have not been observed under practical conditions of food irradiation and thus do not pose specific problems. Methods for the isolation and enumeration of damaged cells from heated or dried foods may be used for these purposes in the examination of irradiated foods, but their applicability should be tested in each case.

No evidence has been reported of enhanced irradiation-induced pathogenicity of foodborne microorganisms, or of increased toxin formation, or induction of antibiotic resistance in irradiated bacteria.
Accordingly, the Committee continues to hold the opinion expressed in 1976 that irradiation of food does not increase the pathogenicity of bacteria, yeasts and viruses.

Because of the intrinsic genetic variability of moulds, experimental results should be interpreted with caution. Laboratory experiments, carried out under conditions which differed greatly from those occurring in practice, have shown that mycotoxin production by moulds derived from irradiated spores may vary (in either direction) in comparison with the parent non-irradiated strain. Other laboratory experiments have shown increased mycotoxin production only if heavy inocula are incubated in irradiated, autoclaved moistened foods. These observations have no relevance to food irradiation under present conditions of practice, in which increased formation of mycotoxins has not been found (see section 8.3.).

6.3. Microbiological aims of food irradiation

It has been demonstrated that irradiation can reduce the

microbial load of a food, thereby increasing the useful life of a perishable food product. The efficacy of irradiation of spices for reducing microbial load is well documented and this process may be a useful alternative to fumigation treatment. Laboratory animal diets have been irradiated successfully for a number of years on a large scale to render them commercially sterile. Salmonella occurs in livestock and is derived from feed and other sources. Since the incidence of such Salmonella can be reduced by irradiation of the feed, this process may afford a means of controlling Salmonella in poultry and some egg products and of dealing with this common public health problem in many parts of the world. The on-shore irradiation of fish and seafood has received much attention because, among other reasons, Vibrio parahaemolyticus is one of the most important foodborne disease agents in warmer climates.

In all, properly designed irradiation processes have been shown to be capable of achieving their intended microbiological objectives (e.g., commercial sterilization, destruction of pathogens). Problems of a microbiological nature that had before been thought might exist have not materialized. Nevertheless, in the case of irradiation, as in any other method of food processing, the gains in microbiological quality must be safeguarded by proper care of the product after processing.

7. TOXICOLOGICAL ASPECTS

7.1. Re-evaluation of provisional acceptances and new evaluations

The Committee reviewed data on fish, onions and rice for re-evaluation and on cocoa beans, dates, mangoes, pulses, and spices and condiments for evaluation. These data were developed in accordance with the guidelines set out in earlier reports of previous Joint Expert Committees. In making its evaluations the Committee used the principles and categories of acceptance, as set out in the previous report (4).

The Committee noted that, in the case of cocoa beans, onions, and spices, the presence of natural constituents exerted toxicologically significant effects when these commodities were fed at high levels in the test diet. These effects were found, whether or not the food had been irradiated. The information available on irradiated vegetables was insufficient to make an evaluation, using the principles previously established. The data on all these commodities were also used in considering the acceptance of irradiated food in general (see section 10).

7.2. Considerations arising from a review of data on irradiated laboratory animal diets and other diets

Concern was expressed by the 1976 Joint Expert Committee about the increasingly common practice of using irradiated prepared feeds for laboratory animals, because of the possible effect on control groups used in toxicological testing (4). Data requested on animal colonies reared on irradiated diets were made available to the present Committee, as summarized below.

Studies comparing diets (sterilized by autoclaving or irradiation at 25-44 kGy or treated to eliminate pathogens at 15 kGy) have been published by institutes in Austria, Denmark, France, Hungary, the Netherlands, and the United Kingdom. These included multigeneration studies in rats (9-14), mice (15-17), and pigs (18). In two of the studies (10, 13), some of the parent and F_1 generation animals were kept for the whole lifespan for information on carcinogenicity. The numbers of animals examined ranged from 5,000 to 500,000.

The Committee concluded from these data that the rearing of test animals on laboratory diets sterilized by irradiation at doses of 15 to 45 kGy was unlikely to obscure any differences if a non-irradiated hygienically acceptable feed had been used.

The Committee also reviewed information on the results of feeding commercial livestock on feedstuffs irradiated at doses of the order of 8 kGy to reduce organisms belonging to the Enterobacteriaceae, especially <u>Salmonella</u>. Breeding and performance studies in poultry (<u>19</u>), and pigs (<u>20</u>, <u>21</u>) produced no evidence to show that feeding of irradiated diet to commercial livestock had any adverse effects.

The Committee was aware of the practice of using totally irradiated diets for maintaining patients on immunosuppressive therapy as the only practical means of supplying palatable food under these conditions. No published systematic investigations or accounts were available to the Committee for evaluation. The absence of reports of adverse effects sugggests that this practice is not deleterious, and this fact was taken into account in the general assessment of the toxicological acceptability of irradiated food. The Committee recommended that if possible there would be a systematic collection and review of information relating to the use of radiation-sterilized human diets.

7.3. <u>Toxicological evaluation of radiolytic products</u>

The Committee reviewed a study in which the principal radiolytic products from irradiated polysaccharides were fed to rats for 6 months at 1700 times the concentration found after irradiation at 3 kGy. No toxic effects were noted (<u>22</u>). These data also support the conclusion set out in section 10. (See also section 4.2.).

8. RE-EVALUATION OF FISH, ONION, AND RICE[1]

8.1. <u>Teleost fish and fish products</u>

Purpose of irradiation

a) To control insect infestation of dried fish during storage and marketing.

---

1. Summaries of the data used in the evaluations and the references are given in a separate document entitled: "Wholesomeness of irradiated food. Summaries of data considered by the Joint FAO/IAEA/WHO Expert Committee, Geneva, 27 October to 3 November 1980". Copies of this document are available, on request, from Division of Environmental Health, World Health Organization, 1211 Geneva 27, Switzerland.

b) To reduce the microbial load of the packaged or unpackaged fish and fish products.

c) To reduce the number of certain pathogenic microorganisms in packaged or unpackaged fish and fish products.

Average dose

For a) up to 1 kGy, and for b) and c) up to 2.2 kGy.

Temperature requirement

During irradiation and storage the fish and fish products
referred to in b) and c) should be kept at the temperature
of melting ice.

Microbiological aspects

Vibro parahaemolyticus is the agent, infectious for man,
that is most typically associated with fish and other seafoods. However, infectious agents derived from the intestines
of man or other warmblooded animals may be present in fish
because these agents were present in the water in which the
fish grew or, as sometimes happens, because they were present
in the only water that was available for cleaning fishing
equipment (including holding compartments on the ship) or
the catch. In addition to infectious agents, toxigenic,
sporeforming bacteria such as Clostridium botulinum type E
may well be present in the fish as caught.

No microbiological problems are likely to arise from
irradiation for purpose a). V. parahaemolyticus will be
eliminated in the product by the doses recommended for
purposes b) and c), while the levels of other pathogens and
spoilage agents will at least be reduced. Irradiation that
does not exceed 2.2 kGy (average dose) is expected to leave
enough spoilage organisms to render the food unacceptable
before cells derived from surviving. C. botulinum spores can
produce enough toxin to constitute a hazard, However, maintenance of the temperature of melting ice throughout the
period of storage of the product has been specified as an
additional safeguard against botulism; salting, drying, or
other effective measures would have to be substituted if this
temperature could not be maintained reliably.

Nutritional aspects

More recent studies have shown that after irradiation at
3 kGy, about 15% of thiamine and 25% of pyridoxine is lost,
while riboflavin, niacin and vitamin B_{12} remain unaffected.
Higher doses confirmed the particular sensitivity of thiamine
and pyridoxine to destruction, the other B complex vitamins
remaining practically unaffected. Further studies have
confirmed the stability to irradiation of the amino-acid
content, particularly of tryptophan. The protein quality of
mackerel and hake remained unaltered even by doses of the
order of 5 kGy.

The lipids extracted from salted dried irradiated mackerel
showed no evidence of adverse nutritional effects at radiation
doses of up to 8 kGy. Irradiation up to a dose of 2.2 kGy
does not appreciably change the usefulness of fish as a good
dietary source of protein, B vitamins, and iodine.

Toxicological aspects

The Committee noted that the results of the studies (on-
going in 1976) had now become available - i.e., short-term,
long-term, reproduction, and dominant lethality studies in
mice; a short-term study in rats, investigating changes in
serum alkaline phosphatase levels when rats were fed on mixed
eviscerated cod and redfish; and short-term and reproduction
studies in rats fed on other fish varieties. These did not
reveal any evidence suggesting that the feeding of irradiated
fish to these animals caused any deleterious effects.

A large number of other feeding studies in which rats and
mice were fed on other varieties of fish and fish products
have also been reported since 1976. These consisted of short-
term and long-term feeding studies and also reproduction,
dominant lethality, and a number of mutagenicity studies.
These new toxicological data, taken together with the results
of previously evaluated studies on various types of irradiated
fish, do not indicate any adverse effects arising from the
administration of irradiated fish to test systems.

Evaluation

The previous provisional acceptance for cod and redfish is changed to unconditional acceptance for fish and fish products irradiated for the purpose of desinfestation, reducing the microbial load, and reducing the number of pathogenic organisms, at an average radiation dose of up to 2.2 kGy.

8.2. Onions
Purpose of irradiation

To inhibit sprouting during storage.

Average dose

Up to 0.15 kGy.

Microbiological aspects

No special microbiological problems of public health significance are known to be associated with irradiated onions.

Nutritional aspects

Recent studies have confirmed the previously reported lack of effect of irradiation, with doses of up to 0.15 kGy, on the ascorbic acid content of onions even after 10 months of storage. The content of reducing sugars increased in irradiated onions to a smaller extent than in untreated onions. No changes occurred in the amino-acid composition.

Toxicological aspects

The requirement of the previous Committee for a multi-generation study in rats, at feeding levels below that causing biological changes due to the biologically active substances that were naturally present, has now been met. In addition, a number of short-term, reproduction, terato-genicity, and dominant lethality studies in rats have now been reported. None of these studies has shown any adverse effects when irradiated onions were incorporated at a 2% level in the diet of rats and mice. Additional corroborative evidence has been obtained from many mutagenicity studies on onions treated (for the prevention of sprouting) with doses of radiation of up to 0.15 kGy and from similar studies on dried onion powder treated with radiation doses of up to 15 kGy.

Evaluation

The previous provisional acceptance is changed to un-conditional acceptance of onions irradiated, for the purpose of controlling sprouting, at an average dose of up to 0.15 kGy.

8.3. Rice
Purpose of irradiation

To control insect infestation in stored rice.

Average dose

Up to 1 kGy.

Prevention of reinfestation

Rice, whether prepackaged or handled in bulk, should be stored as far as possible, under such conditions as will prevent reinfestation.

Microbiological aspects

If the moisture content of stored rice is too high, fungi such as Aspergillus flavus, which are sometimes toxigenic, may grow. Such moulds cannot grow in rice that is stored in a properly dry condition; however, there has been concern over some results that suggested that irradiation could enhance the toxigenic potential of the moulds. It has been shown that toxin-producing fungi are more susceptible than other fungi to irradiation; that a higher water activity is required for the growth of toxin-producing aspergilli than for that of other aspergilli; and that, even at a high water activity, non-toxin-producing strains of Aspergillus overgrow the toxin-producing strains and suppress their formation of toxin. Storage of rice at a sufficiently low level of moisture is critically important; the potential mycotoxin hazard is not enhanced by irradiation under practical conditions.

Nutritional aspects

The loss of thiamine on cooking, noted in the report of

the 1976 Joint Expert Committee (4), may make any further
losses due to irradiation relevant where rice is a staple
item of the diet and a major source of thiamine. However, a
recent study has shown that irradiation at dose levels up to
0.5 kGy did not alter the content of B vitamins or the amino-
acid composition.

Toxicological aspects

   The Committee noted that the results of the long-term
study in rats and the short-term study in monkeys, requested
in 1976 (4), were now available. These showed that the ingestion of irradiated rice caused no adverse effects on the
test animals. Another multi-generation study and a dominant
lethality study in mice, as well as cytogenetic investigations
of the bone marrow of mice and hamsters that had been fed
irradiated rice in their diet, showed no adverse effects.
These additional results, taken together with the results of
the previously reviewed studies, do not indicate any adverse
effects from the ingestion of irradiated rice.

Evaluation

   The previous provisional acceptance is changed to unconditional acceptance of rice irradiated, for the purpose
of controlling insect infestation, at an average dose of up
to 1 kGy.

9. NEW EVALUATIONS[1]

9.1. Cocoa beans
Purpose of irradiation
   a) To control insect infestation in storage.
   b) To reduce the microbial load of fermented beans with
or without heat treatment.

Average dose
   For a) up to 1 kGy, and for b) up to 5 kGy.

---

1. See footnote 1 on page 137.

Prevention of reinfestation

Cocoa beans, whether prepackaged or handled in bulk, should be stored, as far as possible, under conditions that will prevent reinfestation and microbial recontamination.

Microbiological aspects

Members of 11 genera of moulds, some of which are toxigenic, have been found to be natural contaminants of the cocoa bean embryo and are a major factor limiting the storage life of the product. Mould growth flourishes at moisture levels exceeding 8%. Irradiation with doses of 0.5 kGy eliminates moulds in young (under 2 months) beans, whereas a dose of 5 kGy eliminates moulds even in older beans. Pretreatment of cocoa beans with heat (100° C for 10-15 minutes) enhances the radiosensivity of the moulds they contain.

Nutritional aspects

Beans irradiated with doses in the range of 0.1 to 5 kGy showed no significant differences from unirradiated beans with regard to their content of reducing sugars, total amino acids, total fat, and protein. Analysis of cocoa fat in the irradiated material showed no detectable chemical difference from that in unirradiated material.

Toxicological aspects

The available results of the short-term and reproduction studies in rats do not indicate any adverse effect due to the irradiation treatment of the cocoa beans. Both irradiated and unirradiated cocoa beans depressed growth and reduced the food intake when incorporated at high levels in the diet of test animals. The observed toxic effects of the cocoa bean diet on fetal development and survival are related to the high theobromine content of the diet. This has been confirmed by cross-fostering experiments and specific studies using theobromine alone. A number of mutagenicity studies have shown the absence of any mutagenic potential in irradiated cocoa beans.

144

Evaluation

Unconditional acceptance of cocoa beans irradiated, for the purpose of controlling insect infestation or of reducing the microbial load, at an average radiation dose of up to 5 kGy.

9.2. Dates

Purpose of irradiation

To control insect infestation in stored dates.

Average dose

Up to 1 kGy.

Prevention of reinfestation

Prepackaged dried dates should be stored under conditions that will prevent reinfestation.

Microbiological aspects

No microbiological objectives are being pursued by irradiation of dried dates and no public health problems of a micro-biological nature are envisaged.

Nutritional aspects

Irradiation of dried dates with doses in the range of 0.3 to 5 kGy had no effect on the reducing sugar content and on major carbohydrate components. No malonaldehyde was detected. No effect on the protein content was discovered. Irradiation of dates with doses of up to 10 kGy induced no appreciable changes in the amino-acid composition.

Toxicological aspects

The available short-term study in rats revealed no adverse effects that could be related to ingestion of irradiated dates. The results of the reproduction study in rats and of many mutagenicity studies, including a study for induction of recessive lethals in Drosophila, revealed no adverse effects that could be ascribed to the irradiation treatment.

Evaluation

Unconditional acceptance of dates irradiated, for the

purpose of controlling insect infestation, at an average dose of up to 1 kGy.

9.3. Mangoes
Purpose of irradiation
   a) To control insect infestation.

   b) To improve the keeping quality by delaying ripening.

   c) To reduce the microbial load by combining irradiation and heat treatment.

Average dose
   Up to 1 kGy.

Microbiological aspects
   Microbial species isolated from mangoes do not appear to be a threat to human health. Germination of naturally occurring or experimentally inoculated Gloeosporium fusarium and G. singulata is reduced by increasing the doses of irradiation, but complete inhibition requires a dose of 4 kGy, which is technologically unacceptable.

Nutritional aspects
   Several studies have shown that irradiation at dose levels of up to 2 kGy caused only slight losses in ascorbic acid and carotene, compared with the effects of freezing or heat treatment. The contents of riboflavin, niacin, and thiamine are not affected. The levels of fat, protein, sugar, and minerals remain unaffected by irradiation.

Toxicological aspects
   The available investigations included short-term, long-term, multigeneration, and teratogenicity studies in rats as well as a number of mutagenicity studies. The results indicated that the incorporation in the test diets or irradiated mangoes produced no adverse effects.

Evaluation
   Unconditional acceptance of mangoes irradiated for the purpose of controlling insect infestation or for delaying

ripening or reducing the microbial load at an average
radiation dose of up to 1 kGy.

9.4. Pulses

Purpose of irradiation

To control insect infestation in stored pulses.

Average dose

Up to 1 kGy.

Prevention of reinfestation

Pulses, whether prepackaged or handled in bulk, should be
stored, as far as possible, under conditions that will
prevent reinfestation.

Microbiological aspects

No specific microbiological problems arise with pulses,
whether irradiated or not.

Nutritional aspects

Pulses are a major source of dietary protein in certain
parts of the world. Any deleterious effects of irradiation
on the nutritional quality of these crops would therefore
be of importance. Conflicting results appear in studies of
the protein efficiency ratio (PER)[1] and the effects on B-
complex vitamins have not been well established for different
pulses.

These possible effects should receive consideration wherever
irradiated pulses are used as staple items of the diet.

Toxicological aspects

The available short-term studies in mice and rats, as well
as reproduction study in rats, did not indicate any adverse
effects due to irradiation of several varieties of dried

---

1. The protein efficiency ratio is a rough measure of the
nutritive value of proteins, obtained by dividing the gain
in body mass by the mass of the protein consumed. It is
usually measured in young rats, fed on a diet containing 10%
protein under standard conditions.

beans and cowpeas. There was a reduction in the growth rate
of rats after the ingestion of high dietary levels of both
irradiated and unirradiated beans. A number of mutagenicity
studies, including a dominant lethality study in mice, did
not reveal any mutagenic potential in several varieties of
irradiated dried beans.

Evaluation

Unconditional acceptance of pulses irradiated, for controlling insect infestation, at an average radiation dose of up
to 1 kGy.

9.5. Spices and condiments[1]

Purpose of irradiation

a) To control insect infestation.
b) To reduce the microbial load.
c) To reduce the number of pathogenic microorganisms.

Average dose

For a) up to 1 kGy, and for b) and c) up to 10 kGy.

Microbiological aspects

Fungal contaminants, some of which are likely to be toxigenic, occur in untreated spices at an average level of 10^4/g.
Other agents of possible concern to human health include the
food-poisoning species Bacillus cereus and Clostridium
perfringens; Salmonella and Shigella have been reported.
Aerobic spore-formers and thermophilic bacteria at levels of
up to 10^8/g must be dealt with by some means other than heat.
Because the majority of the flora are radiosensitive, irradiation doses of 4-5 kGy reduce the total bacterial counts to
less than 10^4/g. Commercial sterility can be achieved at
doses of 15-20 kGy, depending on the initial microbial load.
The flora that survive irradiation have a lower heat and
salt tolerance, so that the subsequent heat treatment of
products containing the irradiated spices can be reduced.

---

1. Inclusive of "dehydrated onion" and "onion powder".

Nutritional aspects

Irradiation of paprika at temperatures in the range of 0° C to 22° C, with doses of 5-50 kGy, and subsequent storage for 6 months had practically no effect on the carotenoid content.

Radiation treatment with 5 and 15 kGy affected the relative concentrations of some fatty acids but not always in a dose-dependent manner. In some spices there is a small reduction in the proportion of some unsaturated fatty acids. Since spices do not contribute significantly to the nutritional quality of food, these changes are of no nutritional significance.

Toxicological aspects

The available reports of feeding studies in rats (including short-term, reproduction, and teratogenicity studies) are less comprehensive in the case of irradiated spices and condiments than for other irradiated foods. Some of the adverse effects observed in the test animals are related to the ingestion of high dietary levels of spices, both irradiated and unirradiated. No untoward effects, attributable to the irradiation treatment, were reported in these studies. The results of several mutagenicity tests revealed the absence of any mutagenic potential. In evaluating the safety of this commodity, the Committee took into consideration the low levels of spices used in the human diet.

Evaluation

Unconditional acceptance of spices irradiated for the purpose of controlling insect infestation, or of reducing the microbial load and the number of pathogenic microorganisms, at an average radiation dose of up to 10 kGy.

10. CONCLUSIONS ON THE ACCEPTABILITY OF IRRADIATED FOOD

10.1. Toxicological acceptability of irradiated food

The Committee, having reviewed new evidence, was able to formulate a recommendation on the acceptability of food

irradiated up to an overall average dose of 10 kGy (see
sections 2 and 3). This development follows logically from
the approaches to the assessment of the wholesomeness of
irradiated food adopted in the past by previous Joint Expert
Committees, as described in the Introduction. The following
considerations led to this development.

a) All toxicological studies carried out on a large
number of individual foods (from almost every type of food
commodity) have produced no evidence of adverse effects as
a result of irradiation.

b) Radiation chemistry studies have now shown that the
radiolytic products of major food components are identical,
regardless of the food from which they are derived. Moreover,
for major food components, most of these radiolytic products
have also been identified in foods subjected to other,
accepted types of food processing. Knowledge of the nature
and concentration of these radiolytic products indicates that
there is no evidence of a toxicological hazard.

c) Supporting evidence is provided by the absence of any
adverse effects resulting from the feeding of irradiated
diets to laboratory animals, the use of irradiated feeds in
livestock production, and the practice of maintaining
immunologically incompetent patients on irradiated diets.

The Committee therefore concluded that the irradiation of
any food commodity up to an overall average dose of 10 kGy
presents no toxicological hazard; hence, toxicological testing of foods so treated is no longer required.

10.2. Microbiological and nutritional acceptability of irradiated food

The Committee considered that the irradiation of food up
to an overall dose of 10 kGy introduces no special nutritional
or microbiological problems, However, the Committee emphasized
that attention should be given to the significance of any
changes in relation to each particular irradiated food and
to its role in the diet.

10.3. <u>High-dose irradiation</u>

The Committee recognized that higher doses of radiation
were needed for the treatment of certain foods but did not
consider the toxicological evaluation and wholesomeness
assessment of foods so treated because the available data
are insufficient for this purpose. Further studies in this
area are therefore needed.

11. FUTURE RESEARCH

The Committee considered that future research is needed
in the following areas in order to increase existing knowledge about the effects of irradiation on food and to
facilitate future evaluations:
- The technological and economic feasibility of conducting
food irradiation on a larger scale and with a wider variety
of foods should be established (see section 3.).
- Further studies in the area of wholesomeness assessment of
certain foods irradiated at higher doses are desirable (see
section 10.3.).
- If possible, there should be a systematic collection and
review of information on the effects of using irradiation-
treated human diets (see section 7.).
- The conflicting results published on the effect of radiation
on the biological value of proteins and B complex vitamins in
pulses should be clarified because of their importance as
staple foods in many countries (see section 9.4.).
- As there is little recent information on the effect of
radiation on folic acid, future work should be carried out
on representative folate-containing foods, since the diets
in some parts of the world have a marginal folic acid
content (see section 5.).
- Further work on the effects of combination of irradiation
with other processes on the nutritional value of foods so
treated is desirable (see section 5.).

12. RECOMMENDATIONS

The technological and economic feasibility of food irradiation on an industrial scale should be established. A wider variety of foods should also be studied with respect to their suitability for processing by irradiation. IAEA and FAO should facilitate such studies and collect data for the purpose of making recommendations.

The use of high-dose radiation for the treatment of certain foods has been recognized as being technologically feasible. To assess the safety of this process, further information is needed on its nutritional, microbiological and toxicological implications. Such information is being generated and should be brought together by FAO, IAEA and WHO for future evaluation.

REFERENCES

1. Report of the FAO/WHO/IAEA Meeting on the Wholesomeness of irradiated Foods, 23-30 October 1961, Brussels. Rome, Food and Agriculture Organization of the United Nations, 1963.
2. WHO Technical Report Series, No. 316, 1966 (The technical basis for legislation on irradiated food. Report of a Joint FAO/IAEA/WHO Expert Committee).
3. WHO Technical Report Series, No. 451, 1970 (Wholesomeness of irradiated food with special reference to wheat, potatoes and onions. Report of a Joint FAO/IAEA/WHO Expert Committee).
4. WHO Technical Report Series, No. 604, 1977 (Wholesomeness of irradiated food. Report of a Joint FAO/IAEA/WHO Expert Committee).
5. Becker, R.L. A determination of the radioactivity induced in foods as a result of irradiation by electrons of energy between 10 and 16 MeV. US Army Natick Research and Development Command, Contract No. DAAK60-78-R-0007, April 1979. Unpublished report submitted to WHO by the International Food Irradiation Project.
6. Report of a Joint FAO/IAEA/WHO Advisory Group on International Acceptance of Irradiated Foods, Wageningen, Netherlands, 28 November - 1 December 1977. Vienna, International Atomic Energy Agency, 1979 (STI-PUB/530).
7. Evaluation of the health aspects of certain compounds found in irradiated beef. Published by Life Science Research Office, Federation of American Societies for Experimental Biology (FASEB), Rockville Pike, Bethesda, MD, USA, 1977.

8. Evaluation of the health aspects of certain compounds found in irradiated beef. Supplement II. Further toxicological considerations of volatile compounds. Supplement II. Possible radiolytic compounds. Published by Life Science Research Office, Federation of American Societies for Experimental Biology (FASEB), Rockville Pike, Bethesda, MD, USA, 1979.

9. Aravindakshan, M. et al. Multigeneration feeding studies with an irradiated animal feed. In: Use of radiations and radioisotopes in studies of animal production. Proceedings of a Symposium held at the Indian Veterinary Research Institute, Izatnagar. India, 16-18 December 1975. Bombay, Food and Agriculture Committee, Department of Atomic Energy, 1976, pp. 325-332.

10. Aravindakshan, M. et al. Multigeneration feeding studies with an irradiated whole diet. In: Food preservation by irradiation, Vol. II. Proceedings of an International Symposium, Wageningen, Netherlands, 21-25 November 1977. Vienna, International Atomic Energy Agency, 1978. (STI/PUB/470), pp. 41-52.

11. Barna, J. (Wholesomeness test of irradiated complete diet in multigeneration experiment. I. Growth and body weight data (published in Hungarian, summary in English), Budapest, Central Food Research Institute, 1973.

12. Iwado, S. et al. Sterilization of laboratory animal diets by gamma radiation. In: Food irradiation in the Takasaki Radiation Chemistry Research Establishment, No. 1 (April 1964-March 1973). Published by the Japan Atomic Energy Research Institute (Report No. JAERI-M5458), 1973, pp. 34-47.

13. Eriksen, W.H. et al. Comparison of the biological effects in rats of radiation-sterilized and autoclave-sterilized food. Roskilde (Denmark), Danish Atomic Energy Commission, 1973 (Risö Report No. 260).

14. Van Logten, M.J. et al. Investigation of the wholesomeness of autoclaved or irradiated feed in rats. Utrecht, National Institute of Public Health, 1978 (unpublished report No. 33/78 Alg. Tox.).

15. Nadudvary, I. Experience of radiation treatment of laboratory and farm animal feeds in Hungary. In: Decontamination of animal feed by irradiation. Proceedings of an Advisory Group Meeting, Sofia, 17-21 October 1977. Vienna, International Atomic Energy Agency, 1979 (STI/PUB/508), pp. 33-41.

16. Adamiker, D. Practical experiences with irradiation of laboratory animal feed. In: Decontamination of animal feed by irradiation. Proceedings of an Advisory Group Meeting, Sofia, 17-21 October 1977. Vienna, International Atomic Energy Agency, 1979 (STI/PUB/508), pp.113-119.

17. Saint-Lèbe, L. Radicidation et radappertisation des provendes pour rats et souris axéniques hétéroxéniques. Recueil de médicine véterinaire, 155 (10): 805-810 (1979).

18. Sickel, E. Irradiated diet in routine use in conventional-
    ization of gnotobiotic piglets. In: Decontamination of
    animal feed by irradiation. Proceedings of an Advisory
    Group Meeting, Sofia, 17-21 October 1977. Vienna,
    International Atomic Energy Agency, 1979 (STI/PUB/508),
    pp. 133-135.
19. Cox, C. et al. Poultry feed radicidation. 2. Long- and
    short-term poultry feeding trials with irradiated poultry
    feeds. Poultry science, 53: 619-624 (1974).
20. Griese, W. et al. Pasteurization of fish meal by irradia-
    tion. 2. Studies on the harmlessness of feeding fattening
    pigs with fish meal pasteurized by irradiation (in
    German). Zentralblatt für Veterinärmedizin, Reihe B, 23:
    769-778 (1976).
21. Reusse, U. et al. Pasteurization of fish meal by irradiation.
    3. The question of increased rates of mitosis after feeding
    radiation-pasteurized fish meal to pigs (in German).
    Zentralblatt für Veterinärmedizin, Reihe B, 26: 500-509
    (1979).
22. Truhaut, R & Saint-Lèbe, L. Différentes voies d'approche
    pour l'évaluation toxicologique de l'amidon irradié.
    In: Food preservation by irradiation, Vol. II. Proceedings
    of an International Symposium, Wageningen, Netherlands,
    21-25 November 1977. Vienna, International Atomic Energy
    Agency, 1978 (STI/PUB/470), pp. 31-40.

Glossary

FAO	Food and Agricultural Organization of the United Nations, Rome (Italy)
IAEA	International Atomic Energy Agency, Vienna (Austria)
ICRP	International Commission on Radiological Protection
ICRU	International Commission on Radiation Units and Measurements
JECFI	Joint Expert Committee on Food Irradiation
WHO	World Health Organization, Geneva (Switzerland)
Absorbed dose	The absorbed dose is the amount of energy absorbed per unit mass of irradiated matter. The unit of absorbed dose is the gray (Gy); 1 Gy = 100 rad = 1 J/kg
Becquerel (Bq)	Unit of activity, being one radioactive disintegration per second of time. 1 Bq = 2.7027×10^{-11} Ci
Curie (Ci)	Unit of activity, which is being superseded by the becquerel (Bq); 1 Ci = 3.7×10^{10} disintegrations per second = 3.7×10^{10} Bq
Disinfestation	Control of the proliferation of insect and other pests by means of irradiation. (Dose range used is 0.3 to 1 kGy)
D_{min}, D_{max}	Mean minimum and maximum absorbed doses in the product
D_{10}	The required dose to reduce the number of microorganisms by a factor 10 or one log cycle

Gamma radiation	This is an electromagnetic form of energy that is released when Co-60 decays. Gamma rays are very like visible light but with much shorter wavelength
Gray (Gy)	Unit of absorbed dose of ionizing radiation. 1 Gy = 100 rad = 1 J/kg
Isotope	Nuclide having the same atomic number (i.e. the same chemical element) but having a different mass number (i.e. same Z, different A)
Rad	(Radiation absorbed dose), unit of absorbed dose, which is being super-seded by the gray (Gy). 1 rad = 0.01 Gy = 0.01 J/kg = 100 erg/g
Radappertization	The application to foods of doses ionizing radiation sufficient to reduce the number of viable micro-organisms to such an extent that very few, if any, are detectable in the treated food. (Dose range used is 10 to 50 kGy)
Radicidation	Elimination of pathogenic organisms and microorganisms by means of irra-diation. (Dose range used is 0.1 to 8 kGy)
Radurization	The application to foods of doses ionizing radiation sufficient to enhance keeping quality by causing a substantial decrease in numbers of viable specific spoilage micro-organisms. (Dose range used is 0.4 to 10 kGy)

The authors

Prof.Dr. D.A.A. Mossel and Drs. P. van Netten
Department of the Science of Food of Animal Origin
Faculty of Veterinary Medicine
The University of Utrecht
Biltstraat 172
3572 BP UTRECHT
The Netherlands

Prof.Dr. E.H. Kampelmacher
Laboratory of Food Microbiology and Hygiene
Agriculture University
Biotechnion
De Dreyen 12
6703 BC WAGENINGEN
The Netherlands

Ing. D.Is. Langerak
Research Institute ITAL
P.O.Box 48
6700 AA WAGENINGEN
The Netherlands

J.G. Leemhorst
Gammaster
P.O.Box 4250
6710 EG EDE
The Netherlands

Mr. J.Ch. Cornelis
Ministry of Health and Environmental Protection
Radiation Protection Directorate
P.O.Box 5811
2280 HV RIJSWIJK
The Netherlands

Drs. R.M. Ulmann
Dutch Health Council
Food Irradiation Scrutinizing body (Commissie Voedselbestraling)
P.O.Box 95379
2509 CJ THE HAGUE
The Netherlands

Ir. J.G. van Kooij
Head, Food Preservation Section
Joint FAO/IAEA Division of Isotope and Radiation
Applications of Atomic Energy for Food and Agricultural
Development
International Atomic Energy Agency
P.O.Box 100
A-1400 Vienna
Austria

Drs. Frederique Defesche
Young & Rubicam-Koster B.V.
P.O.Box 9220
1006 AE   AMSTERDAM
The Netherlands